IoT Applications, Security Threats, and Countermeasures

Internet of Everything (IoE): Security and Privacy Paradigm

Series Editors:
Vijender Kumar Solanki, Raghvendra Kumar, and Le Hoang Son

Handbook of IoT and Blockchain
Methods, Solutions, and Recent Advancements
Edited by Brojo Kishore Mishra, Sanjay Kumar Kuanar, Sheng-Lung Peng, and Daniel D. Dasig, Jr.

Blockchain Technology
Fundamentals, Applications, and Case Studies
Edited by E Golden Julie, J. Jesu Vedha Nayahi, and Noor Zaman Jhanjhi

Data Security in Internet of Things Based RFID and WSN Systems Applications
Edited by Rohit Sharma, Rajendra Prasad Mahapatra, and Korhan Cengiz

Securing IoT and Big Data
Next Generation Intelligence
Edited by Vijayalakshmi Saravanan, Anpalagan Alagan, T. Poongodi, and Firoz Khan

Distributed Artificial Intelligence
A Modern Approach
Edited by Satya Prakash Yadav, Dharmendra Prasad Mahato, and Nguyen Thi Dieu Linh

Security and Trust Issues in Internet of Things
Blockchain to the Rescue
Edited by Sudhir Kumar Sharma, Bharat Bhushan, and Bhuvan Unhelkar

Internet of Medical Things
Paradigm of Wearable Devices
Edited by Manuel N. Cardona, Vijender Kumar Solanki, and Cecilia García Cena

Integration of WSNs into Internet of Things
A Security Perspective
Edited by Sudhir Kumar Sharma, Bharat Bhushan, Raghvendra Kumar, Aditya Khamparia, and Narayan C. Debnath

IoT Applications, Security Threats, and Countermeasures
Edited by Padmalaya Nayak, Niranjan Ray, and P. Ravichandran

For more information about this series, please visit: https://www.routledge.com/Internet-of-Everything-IoE/book-series/CRCIOESPP

IoT Applications, Security Threats, and Countermeasures

Edited by
Padmalaya Nayak, Niranjan Ray, and P. Ravichandran

CRC Press
Taylor & Francis Group
Boca Raton London New York

CRC Press is an imprint of the
Taylor & Francis Group, an **informa** business

First edition published 2022
by CRC Press
6000 Broken Sound Parkway NW, Suite 300,
Boca Raton, FL 33487-2742

and by CRC Press
2 Park Square, Milton Park, Abingdon, Oxon, OX14 4RN

© 2022 Taylor & Francis Group, LLC
CRC Press is an imprint of Taylor & Francis Group, LLC

Library of Congress Cataloging-in-Publication Data
Names: Nayak, Padmalaya, editor. | Ray, Niranjan K., 1974- editor. |
Ravichandran, P. (Educator), editor.
Title: IoT applications, security threats, and countermeasures / edited by
Padmalaya Nayak, Niranjan Ray, and P. Ravichandran.
Description: First edition. | Boca Raton : CRC Press, 2021. | Series:
Internet of Everything (IoE) : security and privacy paradigm — Includes
bibliographical references and index.
Identifiers: LCCN 2021010117 (print) | LCCN 2021010118 (ebook) |
ISBN 9780367491857 (hardback) | ISBN 9780367643829 (paperback) |
ISBN 9781003124252 (ebook)
Subjects: LCSH: Internet of things–Security measures.
Classification: LCC TK5105.8857 .I678 2021 (print) | LCC TK5105.8857
(ebook) | DDC 005.8/7—dc23
LC record available at https://lccn.loc.gov/2021010117
LC ebook record available at https://lccn.loc.gov/2021010118

ISBN: 978-0-367-49185-7 (hbk)
ISBN: 978-0-367-64382-9 (pbk)
ISBN: 978-1-003-12425-2 (ebk)

Typeset in Times
by codeMantra

Dedication

Dedicated to our beloved students who are the backbones and main pillars of the Nation to carry-forward the latest technology.

Contents

Preface ix

Acknowledgements xi

Editors xiii

Contributors xv

Chapter 1   IoT: The Technological Fad of the Digital Age 1

Corinne Jacqueline Perera

Chapter 2   Significance of Smart Sensors in IoT Applications 15

Vinay Kumar Awaar, Praveen Jugge, and Padmalaya Nayak

Chapter 3   Stochastic Modeling in the Internet of Things 35

Srinivas R. Chakravarthy

Chapter 4   FTLB: An Algorithm for Fault Tolerant Load Balancing
in Fog Computing 65

Ashish Virendra Chandak and Niranjan Kumar Ray

Chapter 5   Real-Time Solar Energy Monitoring Using Internet of Things 87

V. Vijaya Rama Raju and J. Praveen

Chapter 6   Machine Learning Techniques in IoT Applications:
A State of The Art 105

*Laxmi Shaw, Rudra Narayan Sahoo, Hemachandran K.,
and Santosh Kumar Nanda*

Chapter 7   The Farmer's Support System: IoT in Agriculture 119

*Jitendra Kumar Rout, Bhagyashree Mohanty,
Shruti Priya, Pankhuri Mehrotra, and Nidhi Bhattacherjee*

Chapter 8   Development of Intelligent Internet of Things (IoT)-Based
                 System for Smart Agriculture..143

                 Santosh Kumar Nanda, Archana Suresh, Quilo Soman,
                 Debi Prasad Tripathy, and Niranjan Ray

Chapter 9   IoT in Health Care in the Context of COVID-19:
                 An Overview on Design, Challenges, and Application...............163

                 P. Ravichandran and K.K. Venkataraman

Chapter 10   Applications of IoT in Health Care: Challenges and Benefits 177

                 Kedar Nath Sahu, Ravindharan Ethiraj, and
                 Paramananda Jena

Chapter 11   Machine Learning-Based Smart Health-care Systems................195

                 Anil Kumar Swain, Bunil Kumar Balabantaray, and Jitendra
                 Kumar Rout

Chapter 12   An Exhaustive Survey of Privacy and Security Based
                 on IoT Networks..209

                 Santosh Kumar Sahu, Durga Prasad Mohapatra, and
                 Dibya Ranjan Barik

Chapter 13   Distributed Denial-of-Service Attacks in IoT Using Botnet:
                 Recent Trends and Challenges 227

                 Padmalaya Nayak, Surbhi Gupta, and Pallavi Shree

Chapter 14   Detection of Node Cloning Attack in WSN to Secure
                 IoT-based Application: A Systematic Survey..............................247

                 Pinaki Sankar Chatterjee

Index...**261**

Preface

The famous Indian poet Rabindranath Tagore once said, "Most people believe the mind to be a mirror, more or less accurately reflecting the world outside them, not realizing on the contrary that the mind is itself the principal element of creation". As the editors of this book, we have taken his words to heart and moved on in compiling this book with no clear framework for its final form in mind.

Although we had an overall idea for some of the chapters contained herein, the book was formulated, edited, and compiled as the editors spent hours together looking for the best content all through different steps in the process. Many authors sent contributions to us, and it was a very challenging process to decide on the contents of the book.

As editors, we intend to help the readers to understand the wide array of IoT initiatives and projects as it is rapidly changing, highly diverse, and economically emergent in the global context. All the authors are involved in looking for answers and ideas in the IoT fields such as in-depth understanding of IoT, the role of sensors in IoT, cloud storage, agricultural applications, health care applications, and the overall role of Machine Learning in IoT applications. At the same time, we intend to help others to learn how IoT and its application are impacting global learners in many ways. Last but not least, readers of this volume will be better equipped to identify emerging trends, projects, and innovations in IoT as well as new possibilities for research and development. For casual readers, it will also serve as an enhanced understanding of the technology enhancements in terms of educational, cultural, and economic challenges and issues facing various stakeholders in IoT environments. Different chapters will illustrate pressing issues and controversies, where discussion and conflict are presently impassioned. This volume reader will have their own planned objectives. We hope that readers will be inspired to contribute to the prevalent analysis and debate related to IoT and its implementation, regardless of the premise of leafing through various pages of this book.

Acknowledgements

Contributing to a chapter is an individual effort. Many experts from Academia and Industry provided their help to compose this book. We would like to thank all the authors for their valuable contributions and support. Many of the authors have provided their ideas and support while compiling this book. I must acknowledge all the authors, scientists, and professors who helped us to execute this project.

We are honored to publish this book with CRC Press/Taylor & Francis Group. We would like to express our deepest gratitude to everyone who put their tremendous effort to make this project succeed. Our sincere gratitude goes to the Acquiring Editor, Marketing Manager and all those involved in executing this project. In particular, we would like to thank the Series Editor, Dr. Vijender Kumar Solanki, for his help and guidance.

We are grateful to all the reviewers for their valuable time and giving constructive suggestions to refine the book in this correct form. Last but not least, our heartfelt thanks to our Institute's Management for providing excellent support to execute this project.

Editors

Padmalaya Nayak, PhD, is a Professor in the Department of Computer Science Engineering at Gokaraju Rangaraju Institute of Engineering Technology, Hyderabad under Jawaharlal Technological University, Hyderabad, since 2009. She earned her doctoral degree from National Institute of Technology, Tiruchirappalli, India in 2010. Prof. Nayak has 17 years of teaching and research experience in the area of Ad Hoc and Sensor Networks. She has published and presented more than 45 research papers in various international journals and conferences. She has also contributed six book chapters to her credit. She has visited many countries to present her research papers in international conferences. Prof. Nayak has received numerous national-/international awards for her academic contributions toward the education system. She is the reviewer of many IEEE, Springer, and Elsevier journals and conference proceedings. She has worked with several funding projects sponsored by AICTE and UGC. Prof. Nayak was granted one Australian patent and one Indian patent to her credit. She is a member of IEEE, IETE, CSI, and IEANG professional bodies. She is also a member of the advisory committee and technical program committee for several international conferences.

Niranjan Ray, PhD, is an Associate Professor in the School of Computer Engineering, at KIIT University, Bhubaneswar. He earned his PhD in CSE from the National Institute of Technology Rourkela in 2014 and Master's degree from Utkal University Bhubaneswar in 2007. His areas of interest include Mobile Ad Hoc Networks, Wireless Sensor Networks, Internet of Things (IoT) and software-defined network, and fog and edge computing. He is a steering committee member of Odisha IT Society and publication chair of International Conference on Information Technology since 2014. He has published one book on sensor networks and currently serving as an associate editor of *IEEE Consumer Electronics Magazine*. He is on the technical program committee of many international conferences such as IEEE INIS, IEEE ICIIS, and IEEE ICCE. He is listed in *Marquis Who's Who in Science and Engineering, 12th Edition* (2016) and *Who's Who in the World, 32nd Edition* (2015). He is a member of IEEE.

P. Ravichandran, PhD, is a Pedagogical Expert who has two PhDs to his credit: one in Education and the other in Information Technology. His first PhD research title was "Empowering the Teachers in Using Webtools" and the second PhD was titled "Developing a Student Model for Intelligent Tutoring System". He is a life member of AIAER (All India Association for Educational Research). Dr Ravichandran has taught classes for educational doctorate degree students in research and quantitative research methods. He has also published a book on Computer-Based Training, which is used for the diploma students in a Malaysian university. Dr Ravichandran has published 30 papers in international journals and conferences and two book chapters to his credit. Apart from the above academic achievements, he had taken several roles in the field of Education as an author, curriculum designer, module writer, external examiner for PhD, and keynote speaker. He had worked as a research coordinator in various top universities across Malaysia, such as Taylor's University, University of Science Malaysia, and Open University Malaysia by receiving internal research grants. He had also served as a research consultant for Taylor's University, Malaysia in the field of MOOCs and had published many papers at international conferences. Last but not least, Dr Ravichandran taught research methodology courses for students from the University of West Scotland and University of Wolverhampton, UK. During his tenure in leading research methodology classes for postgraduate students, he has also conducted many workshops and in-house training sessions for faculty in research methodology and e-learning. Dr Ravichandran has also conducted e-learning workshops for professors from the Kingdom of Saudi Arabia, and he is also a trained technology-enabled learning trainer for Commonwealth of Learning (CoL). He was assigned as an adjunct lecturer in an American university namely University of the People to teach online master's degree courses since 2009.

Contributors

Vinay Kumar Awaar
GRIET
Hyderabad, India

Padmalaya Nayak
GRIET
Hyderabad, India

Bunil Kumar Balabantaray
Department of Computer Science &
    Engineering
N.I.T. Meghalaya
Shillong, India

Dibya Ranjan Barik
Oil and Natural Gas Corporation
    Limited
Dehradun, India

Nidhi Bhattacherjee
School of Computer Engineering
KIIT Deemed to be University
Bhubaneswar, India

Srinivas R. Chakravarthy
Departments of Industrial and
    Manufacturing Engineering &
    Mathematics
Kettering University
Flint, Michigan, USA

Ashish Virendra Chandak
School of Computer Engineering,
    Kalinga Institute of Industrial
    Technology, Deemed to be University
Bhubaneswar, India

Pinaki Sankar Chatterjee
School of Computer Engineering
Kalinga Institute of Industrial
    Technology
Bhubaneswar, India

Ravindharan Ethiraj
Osmania University
Hyderabad, India

Surbhi Gupta
GRIET
Hyderabad, India

K. Hemachandran
Woxsen School of Business
Woxsen University
Hyderabad, India

Paramananda Jena
Electronics and Radar Development
    Establishment (LRDE), DRDO
Bengaluru, India

Praveen Jugge
GRIET
Hyderabad, India

Pankhuri Mehrotra
School of Computer Engineering
KIIT Deemed to be University
Bhubaneswar, India

Bhagyashree Mohanty
School of Computer Engineering
KIIT Deemed to be University
Bhubaneswar, India

Durga Prasad Mohapatra
National Institute of Technology
Rourkela, India

Santosh Kumar Nanda
Techversant Infotech Pvt Ltd
Trivandrum, India

Corinne Jacqueline Perera
Shangrao Normal University
Shangrao, China

J. Praveen
GRIET
Hyderabad, India

Shruti Priya
School of Computer Engineering
KIIT Deemed to be University
Bhubaneswar, India

Niranjan Kumar Ray
School of Computer Engineering
Kalinga Institute of Industrial
    Technology Deemed to be University
Bhubaneswar, India

Jitendra Kumar Rout
School of Computer Engineering
KIIT Deemed to be University
Bhubaneswar, India

Rudra Narayan Sahoo
TCS
Hyderabad, India

Kedar Nath Sahu
Stanley College of Engineering and
    Technology for Women
Hyderabad, India

Santosh Kumar Sahu
National Institute of Technology
Rourkela, India

Laxmi Shaw
Chaitanya Bharathi Institute of
    Technology
Hyderabad, India

Pallavi Shree
ASET Amity University
Patna, India

Quilo Soman
Techversant Infotech Pvt Ltd
Trivandrum, India

Archana Suresh
Techversant Infotech Pvt Ltd
Trivandrum, India

Anil Kumar Swain
Department of Computer Science &
    Engineering
N.I.T.
Meghalaya, Shillong, India

Debi Prasad Tripathy
Department of Mining Engineering
National Institute of Technology
Rourkela, India

K.K. Venkataraman
Department of ECE, PSG Institute of
    Advanced Studies (PSGIAS)
Coimbatore, India

V. Vijaya Rama Raju
GRIET
Hyderabad, India

1 IoT: The Technological Fad of the Digital Age

Corinne Jacqueline Perera
Shangrao Normal University
Shangrao, China

CONTENTS

1.1   The Technical Fad of IoT 1
1.2   The Evolution of IoT 2
1.3   IoT Hype 3
1.4   IPs in the World of IoT 5
1.5   IPv4 5
1.6   IPv6 6
1.7   Cybersecurity Flags of IoT 6
1.8   Vulnerabilities of IoT Internet Bot 7
1.9   Malware 8
1.10 Concluding Remarks 11
References 12

1.1   THE TECHNICAL FAD OF IoT

This chapter contributes to the dialogue surrounding Internet of Things (IoT) and its prospective impact during the present time of mass digitization. With the ushering of the Fourth Industrial Revolution (Industry 4.0), society has been increasingly immersed in the technical fad of IoT. Industry 4.0 is usually used interchangeably with the notion of the Fourth Industrial Revolution. This global buzz that is connected through mobile and smart digital technologies, is quickly gaining popularity because of its massive real-world applications ranging from consumer and enterprise IoT to manufacturing and industrial IoT (IIoT). For the household segment, consumers use computers and smartphones to remotely take control of their smart homes. Many invest in smart appliances that can be connected to heating appliances, lighting, and electronic devices, to name a few. Similarly, smart buildings use sensors to detect the number of occupants in a given vicinity and auto-adjust the room temperature based on the occupancy. Health care and pharmaceutical facilities also benefit from

1

IoT systems for their inventory management systems. IoT in the agricultural industry and smart farming can monitor weather conditions, plant health, mineral and moisture levels, cattle and poultry health, and more.

1.2   THE EVOLUTION OF IoT

The Internet has been in a constant state of evolution ever since the historical inception of the packet-switching network, Advanced Research Projects Agency Network (ARPANET), created for the American military intelligence in the late 1960s (Kutlu, 2020). Picking up from the turn of the century when things were initially linked to the Internet using radio frequency identification (RFID) tags and operational efficiency took center stage, IoT today has innovatively evolved as an enabler of enhanced living, enticing users with its futuristic appeal through some of the intelligent assistants such as *Google Assistant*, *Apple Siri*, and *Amazon Alexa* (Alieyan et al., 2020; Angrishi, 2017). RFID bourgeoned as the first generation of IoT's tagged things and panned out as the founding technology of IoT. Whitmore et al. (2015) produced documented evidence to suggest that the tracking capabilities of RFID are generally understood to be the precursor of IoT.

Figure 1.1 unravels the evolution of IoT from its days of pre-Internet to the present-day IoT. The pre-Internet 1980s era was the 'Short Message Service' (SMS) or human-to-human (H2H) era where people popularly communicated through SMS. In March 1989, the World Wide Web (www), which is not synonymous with the Internet (Shackelford, 2020), emerged as a virtual platform that used hypertext transfer protocol (HTTP) to transfer files on the web (Kaushik & Tarimala, 2020). By the mid-1990s, the 'www', commonly known as the 'web', grew to be a major communication infrastructure, during the dot-com hysteria; the web was used mostly for email communications or for accessing contextually rich information. The early growth of the web was a vast collection of hypertext markup language (HTML) documents that were hyperlinked (electronically connected) through three types of protocols: HTML, HTTP, and uniform resource locator (URL) (Whitmore et al., 2015). HTML is the language that web pages are written in, while HTTP is the most common protocol, developed specifically for the World Wide Web and was known for its user-friendliness and speed. The URL is the address that locates where any given electronic document resides on the web. Web pages were created when hypertext was combined with the Internet, giving the web browser its name – the World Wide Web.

During the period from 2003 to 2004, Javascript paved the way for the web to be less static and so was created Web 2.0, known as the more interactive and dynamic version of its predecessor, Web 1.0. The phenomenon of Web 2.0 was driven in part by the widespread nascence of blogs, wikis, forums, and e-commerce (Xu et al., 2020; Gross, 2019). Soon thereafter, social media was recognized as a set of Internet-based applications manifested by the popular affordances of Web 2.0, namely, Skype (2003), Facebook (2004), YouTube (2005), and Twitter (2006) (McCarthy et al., 2020; Ghani et al., 2019; McLoughlin & Alam, 2019). Finally, the global economy entered a new wave of growth with the advent of machine-to-machine (M2M) communications that paved IoT's new age of intelligence (Holler et al., 2014; Chen,

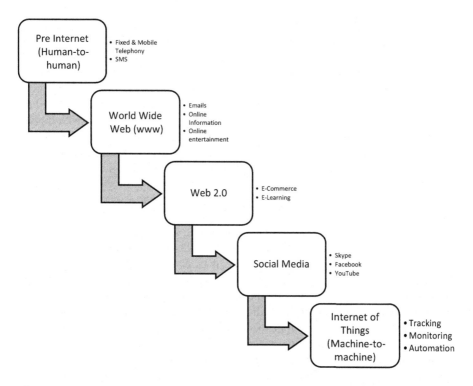

Figure 1.1 Evolution of the Internet of Things.

2012). IoT witnessed the enormous number of potential devices and applications equipped with identification, tracking, monitoring, metering, automation, sensing, and processing capabilities that ultimately enabled these IoT devices to be ubiquitous, context-aware, and ambient intelligent (Whitmore et al., 2015). With the ever-increasing number of smart technologies given greater visibility, the reception of IoT devices has widened people's general perceptions of IoT as an enhanced-living enabler to that of vulnerabilities for cybercriminals (Alieyan et al., 2020). Over time, the widespread adoption of IoT burgeoned out of the advancements made in Internet connectivity, wireless networking technologies, cloud computing, the reduced cost of sensors, memory, and the multitude of connected devices.

1.3 IoT HYPE

The hype of IoT is all about the adoption of IoT technology and its exponential surge. Coined from two words, 'Internet' and 'things', the denotation 'IoT' was coined in 1999 by Kevin Aston, a pioneer on digital innovation. Although the beginning of IoT may be traced back to Aston's RFID experiment and what it originally stood for, nevertheless IoT is in the present day broadly associated with the vast circuitry of gadgets and widgets that leverage on Internet connectivity, operating in real time at an ultra-high-powered gigahertz (GHz) frequency. IoT is a technological paradigm that offers a revolutionary cloud-based infrastructure with an unrestricted constellation of devices, gadgets, electrical appliances, equipment or any form of disparate, physical objects that are coherently connected to the Internet.

The baseline definition of IoT is viewed as a massive web that interconnects physical things of the real world with the virtual world. At any rate, digitalization has made the world more volatile as machines and things are being fitted with digital technology and harnessed for shaping and automating processes. Notwithstanding, it is generally understood that the implementation of IoT is regarded as a plausible solution that helps with the seamless automation of jobs. The integration of IoT, embedded with intelligent sensors and actuators, is essentially crucial in facilitating automation of remotely managed appliances that can enable new ways of production, value creation, and real-time optimization.

On the industrial front, IIoT has proved to be the game changer of industrial automation. IIoT has drawn on the interest of businesses and industry settings with its M2M applications, wireless networks, big data, combination of sensors, artificial intelligence (AI), and analytics that evolved to be the topmost mainstream business growth driver that measured and optimized business processes. As for smart home systems embedded with IoT functionalities, home residents are able to remotely configure and monitor their home appliances such as air-conditioning, light bulbs, TVs, computers, refrigerators, washing machines, and dryers, using either a wall-mounted terminal or a mobile unit that is connected to the Internet and hosted on a cloud infrastructure. Eventually, the IoT hype manifested in homes and offices with smart appliances and devices was taken as a phenomenon that grew relevant to almost anyone.

The five key phases of an IoT technology lifecycle are as follows:

- Innovation trigger
- Peak of inflated expectations
- Trough of disillusionment
- Slope of enlightenment
- Plateau of productivity

Figure 1.2 is a graphical representation of the hype cycle for IoT, which zeros in on the five phases of how emerging technologies and trends evolve throughout its maturity lifecycle. It suggests that every new emerging technology follows through these five phases from its conception to its widespread adoption, representing its perceived value in terms of technological trend or innovation and its relative market promotion.

IoT is about managing tasks from any part of the globe that has Internet connectivity. This connectivity trend poised for growth in the digital age consists of lightweight cryptography (LWC) algorithm developed for smart devices. IoT is referred to as interoperable, smart devices connected through RFID, Bluetooth, QR codes, sensor networks, and wireless technologies (Alieyan et al., 2020). The sensors or actuators embedded within these 'things', add a level of digital intelligence to these 'things'

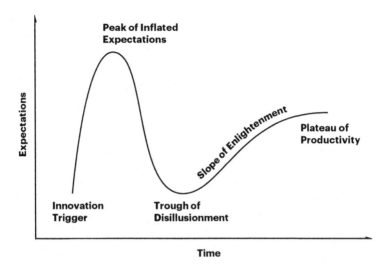

Figure 1.2 IoT hype cycle.

that would otherwise be dumb, enabling them to act in unison and transmit real-time data through an Internet protocol (IP) address. Typically, the IoT value chain is underpinned by IP-based networks that transmit a very large amount of data in real time. The IoT model evolved rapidly due to the progress and enhancements made in its IP network connectivity over a highly scalable bandwidth range.

1.4 IPs IN THE WORLD OF IoT

In the previous two decades, a robust suite of protocols and standards were designed to cope with the web boom and the massive surge of IoT-enabled devices, applications, and Internet growth. Any IoT-enabled device is linked to the Internet through an IP address. The IP is a technology enabler that supports a humongous number of Internet-connected devices from industrial sensors and home appliances to vehicles. IPs are essentially web addresses assigned to Internet-based devices, sites, and services. There are two types of IP addresses: (i) IP version 4 (IPv4) and (ii) IP version 6 (IPv6).

1.5 IPv4

The IPv4 standard is the basis of web addresses. It consists of a unique number based on the combinations of four integers, which amounts to approximately 4 billion possible combinations. The legacy IPv4 with its 32-bit address field is known to provide a maximum of 2^{32} or 4.2 billion Internet-enabled devices (Ziegler et al., 2015; Dooley & Rooney, 2013). The massive surge and proliferation of IoT-enabled devices are depleting the number of available IP addresses on IPv4. With the finite number of IP addresses that are being depleted at an intensifying rate, the significant growth of IoT that requires end-to-end Internet connectivity will certainly not be able to rely on the legacy IPv4 in the coming decades. The Internet is running out of room space. Fortunately, the shortfall of IPv4 has been salvaged by IPv6 in February of 2011, quadrupling the number of IP addresses for the benefit of all users.

1.6 IPv6

IPv6 is the enhanced protocol standard designed to handle the wide spectrum of heterogeneous IoT-based devices. This version offers a unique 128-bit IP address for each connected device. Its 128-bit address field provides 2^{128} or 340 trillion, trillion, trillion, or 340 undecillion (3.4×10^{38}) IP addresses (Ziegler et al., 2015). This is a tremendously, humongous expansion of the quantity and size of the IP address field, which works out to be four times the number of address bits in IPv4 (from 32 bits) to 128 bits (in IPv6). The format of version 6 is hexadecimal and is made up of eight pairs of octets separated by colons. The IPv6 emerged essentially as a tunneling mechanism that transfers data packets using Ping v6 (Packet Internet Groper) and Trace route v6 functions to verify the connections between networks and network devices. Other significant aspects of IPv6 include the essentials of error-reporting, route discovery, diagnostics, and auto-configuration, which allow devices to auto-determine their own IPv6 addresses (Allied Telesis, 2017).

IPv6 offers important benefits. The massive scalability of IPv6 and its current rate of uptake can support up to an enormous number of IP addresses provisioned for the billions of devices, web services, and online applications that favor its adoption and allow IoT to reach its potential. Besides offering solutions for the eventual depletion of IPv4 address spaces, the adoption of IPv6 clearly proves that it can prevent the address apocalypse that was predicted.

1.7 CYBERSECURITY FLAGS OF IoT

The massive growth of interconnected IoT devices poses an enormous range of cybersecurity challenges because of the globally pervasive influence of Internet technology and its increasing adoption in online communications and e-transactions. As IoT devices are being deployed in record numbers, the rise in their popularity comes with its fair share of potential challenges and security vulnerabilities. Some of the pressing IoT security concerns include the following:

- Ransomware attacks and hijacking user data.
- Insufficient testing and lack of timely software updates.
- Hackers gaining unlawful access to users' IP addresses.
- Identity fraud and financial crime.
- Remote smart vehicle access.
- Installing counterfeit IoT devices in secured networks.
- Lack of awareness among users.

Poorly secured IoT devices stand the risk of being exploited by distributed denial of service (DDoS) attacks. DDoS attacks expose security vulnerabilities and cause service disruptions; this can adversely impact consumers' opinions, business productivity, and profits.

1.8   VULNERABILITIES OF IoT INTERNET BOT

The Internet bot is a collection of compromised IoT devices that have been infected with malware. They are typically known under various labels such as Internet robot, web robot, spider, crawler, or zombie. By definition, an Internet bot is a malicious software that is manipulated remotely by a botnet attacker (also known as bot herder or bot master), by remotely controlling the end user's infected computer, through a command and control (CC) network protocol communication channel between bots and server. The bot master commands and controls these IoT devices remotely to execute malevolent activities through a brute force login or by injecting malware through an open port or vulnerable service.

By the same token, cybercrimes are malicious attempts made by cybercriminals who remotely hijack zombie computers (network of infected computers) to steal secure and private information, send spam (used in DDoS extortion scams), or run arbitrary network services for phishing. The evolution of cybercrime is tied to financial gain. Much in the same way, botnets (short for bot network) are a huge number of zombie computers infected with bot malware, such as computer viruses, key loggers, or other malicious software, while being controlled remotely by cybercriminals, usually for their ultimate financial gain. The threat of IoT botnets will continue to increase with the rise of IoT devices deployed. The vulnerability of these devices will always remain an attractive target and attack vector of Internet bots. Besides, it is a known fact that IoT devices are easier to be attacked than regular computers (laptops or desktops). Moreover, in a recent study, Nguyen et al. (2020) reported the following susceptible functionalities and common constraints of IoT devices with regard to Internet bot attacks:

- IoT devices run 24/7 without security updates.
- Usually, IoT devices use the manufacturer's default login credentials or some weak predictable username and password.
- Approximately 20% of the tested routers expose their telnet port to the Internet, which is considered a serious security overbearing.

The concept of bots is commonly associated with cyberterrorism whereby a single Internet bot can be configured to send out a destructive instruction to a targeted CC server, with the intention of disrupting its network activities and compromising the server. Notwithstanding, the advent of IoT botnets, the ubiquity of IoT, and the high vulnerability of IoT devices have become attack vectors for hackers, who are fixated on orchestrating distributed denial-of-service (DDoS) cyberattacks. It has been documented that 96% of devices involved in DDoS attacks are IoT devices (Makhdoom

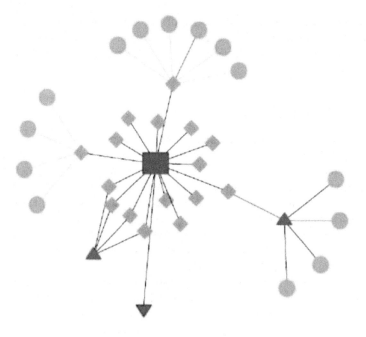

Figure 1.3 Botnet structure.

et al., 2018). According to Gummadi et al. (2009), malicious bot activity is the orig-inator of email spams, DDoS, and 'click-fraud' on advertisement-based websites. With such organized cybercrimes, botnets can always gain anonymous access to the Internet, leaving the hackers untraceable. Figure 1.3 illustrates a botnet structure and how a DDoS attack is triggered off. The inverted triangles represent the bot herder; the large red rectangle in the center represents the CC server; and the yellow squares are bots that are commanded by the bot herder to identify new cyberattack victims.

Figure 1.4 illustrates a typical bot attack flowchart. It describes the workflow pro-cess in which the bot master issues an attack command to the CC server, which then activates the nodes inside the botnet to execute the desired attack.

1.9 MALWARE

One of the prominent security threats that the Internet faces in current times is *mal-ware* (a portmanteau for malicious software). Malware is a blanket term for harmful software programs designed to infiltrate computers and cause harm to user's com-puters without the user's knowledge. Malware comes in a variety of deceptive forms, programmed to execute harmful and destructive cyberattacks. Table 1.1 tabulates the list of malicious software which gain unauthorized intrusion into targeted systems, with the sole purpose of committing cyber theft, sabotage, or espionage. Two of the more popular types of malware (phishing and spyware) are described in further detail below.

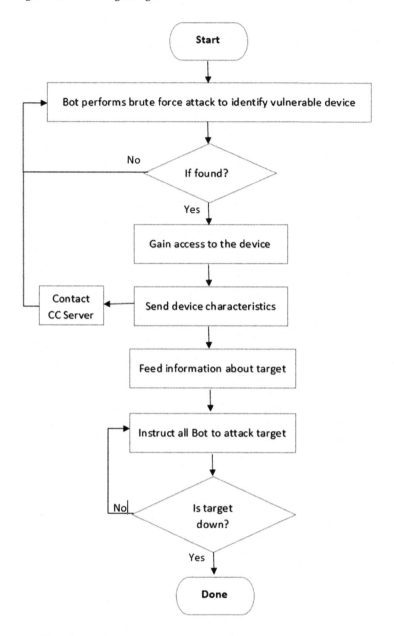

Figure 1.4   Flowchart of bot attack.

Phishing is a malware that is designed to mimic a real app; it can masquerade and pose like a real, trustworthy app on a user's device. A phishing scam is a fraudulent attempt of stealing confidential details of users such as usernames, passwords, social security numbers, driving license credentials, and credit card information. Spam

Table 1.1

Malware

Type	Function	Real-World Examples
Fileless Malware	Distorts files that are native to the Operating System	Astaroth
Spyware	Tracks user's surfing activity without their consent	DarkHotel
Adware	Pop-ups advertisements meant to redirect users to advertising sites and alter browser settings	Fireball
Trojans	Attack routers on wireless networks	Emotet
Worms	Propagate over computer networks to infect other computers.	Stuxnet
Rootkits	Give hackers remote control of a victim's device	Zacinlo
Keyloggers	Designed to track users' keystrokes and capture targeted passwords, financial data, and more.	Olympic Vision
Bots	Self-propagating malware that performs automated commands that are capable of launching floods of DDoS attacks.	Echobot
Mobile Malware	Infects targeted mobile devices.	Triada

emails bearing falsified corporate trademarks, like logos and corporate websites, are sent to prospective victims, alerting them of some sort of problem with their account. Out of desperation, these victims, who are usually online banking users or credit card customers, fall prey to these phishing scams when it triggers off their immediate response towards disclosing their personal and financial data.

Another form of malware called *Spyware* actually spies on users. Cookies are also very identical to *Spyware*, in that it tracks the browsing history on devices and on the data being sent and received. *Spyware* is often attached to free online software downloads or to links that are clicked by users. *Spyware* can secretly monitor an unsuspecting user's computer and steal sensitive personal data, like snagging passwords or stealing confidential files. Therefore, anti-spyware needs to be installed. For computer viruses, a suitable anti-virus should be installed to ward off computer viruses; however, anti-virus scans do not always detect spyware. In general, *Spyware* attacks may be prevented by observing the following precautionary security measures:

- Ensure that anti-virus and anti-spyware updates and patches are maintained regularly.
- Ensure that all downloads are from well-known and reputable sites.
- A firewall is recommended for enhanced security.

Malware may be installed manually either through physical access to a victim's computer or through online remote administrator's access. They may take the form of programs or codes that are intentionally intrusive and malicious. Malware may also be configured to steal, encrypt, or delete data. It can also invade, hijack, or disable the core operating functions of any Internet-enabled device. A malware infection is a serious threat that can lead to things like identity theft and credit card fraud. It can operate without being detected. Oftentimes, smartphone users are prone to malware attacks when they download different apps for various purposes such as social networking, online gaming, and photography. According to Ahvanooey et al. (2020), the younger populace of smartphone users are generally not overly concerned about whether their downloaded apps are infected by malware. However, this study reports of an incident that took place in 2016 whereby a virus, 'HummingBad', had infected over 85 million Android smartphones around the world.

The anatomy of malware attacks stems from the vulnerability of IoT devices that have been viciously manipulated into perpetrating identity theft. According to documented evidence, there have been numerous malicious apps that have stolen users' information from their IoT devices, such as Mirai, Ransomware, Shamoon-2, and DuQu-2 (Makhdoom et al., 2018; HaddadPajouh, et al., 2018). Over time, this has led to huge losses for organizations in terms of revenue generation, brand niche, customers' trust, business secrets, and other such vital statistics. The two topmost entry gateways for a malware access are through the Internet and email. Realizing that all IoT that are connected to the Internet are predominantly malware targets, anything downloaded through the Internet onto a device that lacks the necessary anti-malware security, may be vulnerable to being hijacked by cyber-hackers using a malware to penetrate one's virtual domain. These can also include surfing hacked websites, downloading infected files, clicking on game demos, installing new toolbars from unfamiliar online sites, or opening a *malspam* (malicious email attachment). Some of the possible red flag forewarnings of a malware attack include the following:

- Frequent system crashes and system freeze that display fatal error messages;
- Monitor displays a 'blue screen of death' (BSOD).
- Operating system that slows down in speed.
- Tremendous loss of disk space caused by malware residing in hard drive.
- Pop-up ads that hide malware threats.
- New plugins, toolbars, or extensions that unexpectedly begin populating the Internet browser.

Year in, year out, billions of dollars are being lost to online scams and identity theft. In response to the huge threats posed by IoT devices in inducing identity theft, Weisman (2014) advices against downloading virus-laden malware usually found in free music downloads, online gaming, and videos; as a precautionary security measure against malware attacks, this study recommends the use of an isolated computer, reserved strictly for Internet banking transactions and online purchases.

1.10 CONCLUDING REMARKS

As we enter the dawn of Industry 4.0, IoT technology has become a conversational topic for both household and corporate consumers. With IoT merging the digital and physical universes, billions of Internet-enabled devices embedded with sensors have been raised to a level of digital intelligence that makes them smarter and responsive to real-time communication, independent of any human intervention. This chapter

presents an overview of contemporary evidence related to the technological fad of
IoT, with a view of tracing its historical aspect, assigning IP addresses to Internet-
enabled devices, identifying cybersecurity vulnerabilities, highlighting potential solutions, and proposing future directions in line with emerging technologies. The underlying concept of IoT provides the reader with a comprehensive understanding of
this phenomenon, its current state, emergent technologies, applications, challenges,
and recent developments, all of which provide a non-partisan viewpoint of IoT and
its security domains.

IoT has indeed grown exponentially as the fastest growing technology of various
domains from both industrial and consumers' perspectives. The rapidly evolving IoT
industry has taken a generational leap through its telecommunications value chain by
means of sensors, actuators, and process automation. As a result, the rapid proliferation of the Internet has witnessed the robustness of IoT's wireless technology, including its accessibility of cloud computing, fog computing, and big data processing and
analytics. It is indisputable that IoT has its fair share of impact in many applications
that affect the lives of home users and workplace productivity, which include wearables, cars, homes, cities, and industrials. The tech analyst company IDC predicted
that utilities will be the highest user of IoT because of the advent of smart users,
followed by intruder detection and web cameras, being the second biggest use of IoT
devices. These may be considered the prelude to the future Internet environment that
has emerged as a trend of ubiquitous Internet technology. Presently, IoT continues
to evolve at a breakneck pace catalyzing a technological revolution that represents
the future of the Internet within its ecosystem. As the key insights and benefits of
IoT are slated to improve end user's experience, engagement, and satisfaction, true
market leaders will be the first ones to recognize that devices or 'things' are not the
end game in the IoT ecosystem, but rather to harness on the IoT panacea aimed at
ensuring operational speed and efficiency of vast volumes of data that support their
business needs much more intelligently than their competitors.

REFERENCES

Ahvanooey, M. T., Li, Q., Rabbani, M., & Rajput, A. R. (2020). A survey on smartphones security: Software vulnerabilities, malware, and attacks. arXiv preprint
arXiv:2001.09406.
Allied Telesis. (2017). Technical Guide. Internet Protocol v6 (IPv6): Feature
Overview and Configuration Guide. Retrieved from https://www.alliedtelesis.com
/sites/default/files/ipv6_feature_config_guide_rev_b.pdf.
Alieyan, K., Almomani, A., Abdullah, R., Almutairi, B., & Alauthman, M. (2020). Bot-
net and internet of things (IoTs): A definition, taxonomy, challenges, and future
directions. In B. B. Gupta and R. C. Joshi (eds.), *Security, Privacy, and Forensics
Issues in Big Data*, 304–316. IGI Global, Hershey, PA.
Angrishi, K. (2017). Turning internet of things (IoT) into internet of vulnerabilities
(IOV): IoT botnets. arXiv preprint arXiv:1702.03681.
Chen, Y. K. (2012). Challenges and opportunities of internet of things. In *17th Asia and
South Pacific Design Automation Conference*. January, 2012. IEEE, 383–388.

Dooley, M., & Rooney, T. (2013). *IPv6 Deployment and Management*, Vol. 22. John Wiley & Sons, Hoboken, NJ.

Ghani, N. A., Hamid, S., Hashem, I. A. T., & Ahmed, E. (2019). Social media big data analytics: A survey. *Computers in Human Behavior*, 101, 417–428.

Gross Jr., W. F. (2019). Monitoring and tracking ISIS on the Dark Web. Online Terrorist Propaganda, Recruitment, and Radicalization, 341.

Gummadi, R., Balakrishnan, H., Maniatis, P., & Ratnasamy, S. (2009, April). Not-a-Bot: improving service availability in the face of botnet attacks. In *6th USENIX Symposium on Networked Systems Design and Implementation (NSDI)*, Vol. 9, 307–320.

Haddad Pajouh, H., Dehghantanha, A., Khayami, R., & Choo, K. K. R. (2018). A deep Recurrent neural network based approach for internet of things malware threat hunting. *Future Generation Computer Systems*, 85, 88–96.

Holler, J., Tsiatsis, V., Mulligan, C., Karnouskos, S., Avesand, S., & Boyle, D. (2014). *Internet of Things*. Academic Press, Cambridge, MA.

Kaushik, A., & Tarimala, N. K. R. (2020). U.S. Patent No. 10,645,585. Washington, DC: U.S. Patent and Trademark Office.

Kutlu, Z. G. (2020). Industry 4.0 and the Internet of Things (IoT). In *Internet of Things (IoT) Applications for Enterprise Productivity*, 1–24. IGI Global.

Makhdoom, I., Abolhasan, M., Lipman, J., Liu, R. P., & Ni, W. (2018). Anatomy of threats to the internet of things. *IEEE Communications Surveys & Tutorials*, 21(2), 1636–1675.

McCarthy, P. X., Rizoiu, M. A., Eghbal, S., & Falster, D. S. (2020). Long-term trends of diversity online. arXiv preprint arXiv:2003.07049.

McLoughlin, C. E., & Alam, S. L. (2019). A case study of instructor scaffolding using Web 2.0 tools to teach social informatics. *Journal of Information Systems Education*, 25(2), 4.

Nguyen, H. T., Ngo, Q. D., Nguyen, D. H., & Le, V. H. (2020). PSI-rooted subgraph: A novel feature for IoT botnet detection using classifier algorithms. *ICT Express*, 6, 128–138.

Shackelford, S. J. (2020). *The Internet of Things: What Everyone Needs to Know®*. Oxford University Press, Oxford.

Weisman, S. (2014). *Identity Theft Alert: 10 Rules You Must Follow to Protect Yourself from America's #1 Crime*. FT Press, Upper Saddle River, NJ.

Whitmore, A., Agarwal, A., & Da Xu, L. (2015). The Internet of Things—A survey of topics and trends. *Information Systems Frontiers*, 17(2), 261–274.

Xu, F., Pan, Z., & Xia, R. (2020). E-commerce product review sentiment classification based on a naïve Bayes continuous learning framework. *Information Processing & Management*, 57, 102221.

Ziegler, S., Kirstein, P., Ladid, L., Skarmeta, A., & Jara, A. (2015). The Case for IPv6 as an Enabler of the Internet of Things. IEEE Internet of Things.

2 Significance of Smart Sensors in IoT Applications

Vinay Kumar Awaar, Praveen Jugge,
and Padmalaya Nayak
GRIET
Hyderabad, India

CONTENTS

2.1 Introduction 15
2.2 The Job of a Sensor 16
    2.2.1 Sensors vs. Smart Sensors 17
    2.2.2 Smart Sensors Overview and Technology 20
    2.2.3 Micro-Electro-Mechanical Systems Technology 20
    2.2.4 Advantages of Smart Sensors 21
2.3 Smart Sensors in IoT Applications 21
    2.3.1 List of Smart Sensors Used in IoT Applications 22
    2.3.2 Classification of Smart Sensors 23
    2.3.3 Sensor Materials 23
2.4 Role of Sensor Networks in Various Applications 24
    2.4.1 Smart Home Application – A Case Study
        (Our Developed Model) 28
    2.4.2 Challenges and Issues in Developing Smart Homes 30
2.5 Conclusion 31
References 32

2.1 INTRODUCTION

Internet of Things (IoT) is considered as a key technology in the present world scenario. The infrastructure of IoT poses an ecosystem consisting of people, process, and technology and uses smart devices with embedded systems, cloud storages, complete Internet structure, and applications needed by the users (Khan and Yuce, 2019). IoT-based technology has made our life smoother and easier. It has given a shape to both the industrial and consumer worlds. With the exponential growth of applications and the increased number of connected devices, severe challenges have arisen (Colakovi¢ and Hadziali¢, 2018). The applications of IoT have shown their potentials in various fields such as health care, agriculture, logistics, supply chains, education, and many more (Nayak, 2016). Moreover, to build up an IoT infrastructure, two things are more important: the Internet and physical devices like sensors and actuators. Different types of applications require different types of sensors to collect data from the environment. The bottom layer of the IoT system consists of sensor connectivity and a network to collect information. This layer is an essential part of the IoT system and has network connectivity to the next layer, which is the gateway and network layer. Smart sensors sense and record the data accurately and maintain the collection of environmental data automatically with a minimum error (Safak, 2014). These sensors are used in many applications such as smart grids, spaceships, satellites, military equipment, and a variety of science and engineering applications. The capability of the IoT increased with the integration of smart sensors and becoming an integral part of IoT. Motivated by this fact, we aim to study and discuss smart sensors in detail as well as highlight their applications through a case study.

The remaining part of the chapter is structured as follows. Section 2.2 discusses the job of a sensor and explains the difference between sensors and smart sensors. Section 2.3 discusses different materials used in smart sensors and roles of smart sensors in IoT applications. Section 2.4 discusses the sensor networks applications and presents a case study of smart home control system followed by a conclusive remark in Section 2.5.

2.2   THE JOB OF A SENSOR

The main job of a sensor is to sense the data from the surrounding environment and send it to a processor, where it can be further processed for a meaningful application. Sensors or 'things' represent the front end of the IoT system. These are connected to IoT networks directly or indirectly after signal conversion and processing. However, all the sensors are of no same type and made up of different materials. Therefore, different types of sensors are used for different applications. For instance, digital sensors are easy to interface with a microcontroller using the serial peripheral interface (SPI) bus, but for analogue sensors, either analogue-to-digital converter (ADC) or sigma-delta-modulator is used to convert the data into SPI output. There are many types of sensors available. The cost of a sensor ranges from 10$ to 1000$. Various types of sensors are depicted in Figure 2.1. For instance, temperature sensors, humidity sensors, gas sensors, motion sensors, smoke sensors, and accelerometer sensors are used to measure the physical properties of the environments. Infrared sensors or proximity sensors are used in various applications to detect an object within proximity. Almost all mobile phones carry IR sensors, like proximity sensors, to detect an object. Similarly, ultrasonic sensors are used to measure the distance and velocity of an object.

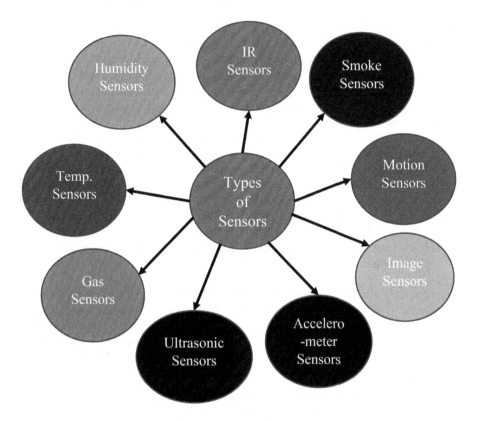

Figure 2.1 Types of sensors.

2.2.1 SENSORS VS. SMART SENSORS

There is a primary difference between a sensor and a smart sensor. A sensor senses and sends the data for remote processing after receiving a specific input signal, whereas an intelligent sensor does the computational processing before sending the data for the remote operation (Pirbhulal et al., 2017). A primary sensor contains transducers, analog filters, amplifiers, compensation devices, and control algorithms, whereas a smart sensor comprises many other components like software-defined elements to perform digital processing, digital filtering, communication processing, data conversions, and so on. The definition of the term "sensor" and its interpretation is different in many dictionaries and research articles. According to IEEE std. 1999, "sensor" is a component that provides a useful output in response to a physical, chemical, or biological phenomenon. This component may already have the same signal conditioning associated with it. Subsequently, as per the IEEE std. 2003, the term "sensor" is defined as a "transducer that connects a physical, biological, or chemical parameter into an electrical signal". It is also included in the IEEE std. 1999 that the term "transducer" is a "device converting energy from one domain to another, calibrated to minimize the errors in the conversion process". The definition of "senor" is huge in range, and a quick presentation shows that the sensors can transform the signal from one domain to another as depicted in Figure 2.2 (Liu and Baiocchi, 2016). Hence, in the essence of many standards and research articles, the term "sensor" can be interpreted as a remote device or a part of a complex device. However, in practice, the "sensor" is designated in different ways. It can be characterized as follows:

- A sensitive device
- It contains a group of sensitive elements to work as a primary measuring transducer. This group of sensitive elements can be connected in a series that makes it a measuring transducer.
- It is an isolated device that can be any of the device mentioned above and, in any combination.
- It contains additional components for signal processing such as ADC, MCU, and so on.

The signal domains of a sensor can be converted into electrical signals, which require an accurate transducer to get the actual signal conversion. For instance, a mechanical domain signal from accelerometer can be converted into electrical signals to measure the acceleration of the structure in three dimensions. Since the piezoelectric material is squeezed by the force caused by the change in motion, it thus produces

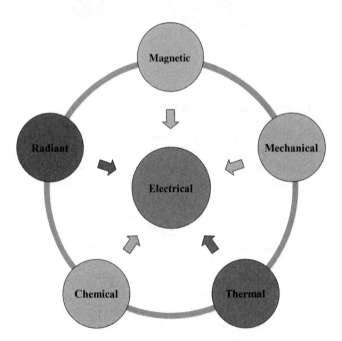

Figure 2.2 Classification of sensors based on signal domains.

the electrical charge that is proportional to the force exerted upon it. Therefore, the piezoelectric material works as a transducer that converts the electrical signal from the pressure, which must be accurate to get the proper signal generation. Likewise, there are many transducers are available for various signal conversions (Gubbi et al., 2013).

For instance, an image sensor for the radiant domain; the temperature sensor for the thermal domain; a hall sensor for the magnetic domain; a PH sensor for the chemical domain, and the electrical domain needs signal conversion from one electrical signal to another electrical signal. Even though sensors operate the data received by the transducer and send the data for further process, the erroneous data must be controlled or corrected before sending it to the signal processing. Thus, it requires an additional integration of a device that can generate a correct signal and sends it to the controlled network environment, which makes a sensor as a "smart sensor" (Boubiche et al., 2018).

Smart sensors: According to the IEEE std. 1451.2 (1997, 1999, 2003), a "smart sensor" is a "sensor version of a smart transducer". Subsequently, a smart transducer is a "transducer that provides the functions over and above that necessary for generating a correct representation of a sensed or controlled physical quantity" (Geirinhas, 2007). A smart sensor's functionality is to integrate the transducer into applications of the controlled network environment. However, the additional functions of the smart sensors can differ from application to application and customer to customer. A smart sensor is characterized with the following additional features of a sensor: (i) self-check and (ii) self-identification.

Therefore, a "smart sensor" is a single-chip device without any external component but includes sensing functionality. It has an interface and is capable of performing signal processing with the intelligence that makes it a self-check and self-validating device. Furthermore, all the functions can be integrated into a single chip. A smart sensor is a combination of an analogue interface circuit, ADC, a memory unit, a controller, and a bus interface with a microcontroller unit (MCU), in addition to the sensor element as shown in Figure 2.3.

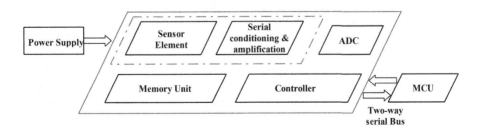

Figure 2.3 Smart sensor system.

2.2.2 SMART SENSORS OVERVIEW AND TECHNOLOGY

The ability to sense the signal domains, as described in Figure 2.2, such as magnetic, mechanical, thermal, electrical, chemical, and radiant by the sensors can be made possible with the combination of semiconductor processing and the micromachining technology or microsystem technology (MST). A subset of this MST is known as micro-electro-mechanical systems (MEMS) (Gad-el-Hak, 2005). However, these MEMS sensing techniques can be varied for each stimulus of the sensor. For instance, the measurement of a parameter considerably varies depending upon the requirements of the accuracy, the range of the actuator, environmental considerations, the signal characteristics, and other inputs of the actuator. The primary environmental considerations can be the sensor temperature, the chemical conditions where the sensors are exposed, and the media of the data communication. Apart from these parameters, certain constraints would affect system performance such as the data transmission, signal transition, signal conditioning, the data display, life span of the device, the unit calibration, impedance of the system/sensor, the source voltage, frequency response, time delay, noise filtering methods, and so on. These constraints must be considered while providing a solution to an existing measuring method. In some instances, sensing technologies can be integrated with more than one technology to make the sensor's design more flexible. However, such a condition depends on the requirements of the applications and the capability of the sensor manufacturers/suppliers.

For instance, the measurement of velocity can be done by the hall-effect sensor, inductive sensor, optoelectronic sensor, and magneto-resistive sensor. In addition, these sensors can be used to measure displacement. Thus, for a specific application using semiconductor sensors, it is quite possible to integrate piezoresistive or capacitive technology. However, once a specific technology is adopted for a measurement, it is hard to choose another technology as an alternative. It is imperative to understand the fundamentals of the design principles of the sensor, along with the specifications to determine and select an effective sensing technology.

2.2.3 MICRO-ELECTRO-MECHANICAL SYSTEMS TECHNOLOGY

There are two essential constructional technologies of microengineering. One is *microelectronics*, and the other is *micromachining*. The technology used for producing electronic circuits on silicon chips is called microelectronics. In contrast, the techniques used for moving parts and structures of micro-engineered components are called micromachining. These two technologies can be integrated to produce microsystems, i.e., to produce a microelectronic circuitry into a micromachined structure. Such systems have a few advantages such as small size, low cost, and reliability. Nowadays, three micromachining technologies are extensively used by the manufacturing industries. These are as follows:

- **Silicon micromachining**: This technology has become the most prominent technology among the three techniques since silicon is the primary substrate material used for microelectronic circuitry production. Eventually, silicon is the preferable material in the production procedure of microsystems. The conventional laser methods burn or vaporize the materials for extracting them, which may damage the sensitive materials.

- **Excimer laser:** Unlike these methods, Excimer laser uses an ultraviolet laser technique to micromachine (several materials) without heating them. This technology is best suited for organic materials like polymers.
- **LIGA:** LIGA is an acronym in German for the process of Lithographie, Abformung, and Galvanoformung (LIGA). The technique LIGA consists are lithography, melding, and electroplating processes for the microstructures production.

2.2.4   ADVANTAGES OF SMART SENSORS

MEMS technologies have made the smart sensors more compact and flexible to many applications in recent times, like in process industries, IoT, medical, and many more areas where accuracy is achieved through direct digital control. With the smart sensors, it is possible to have conditional monitoring of the devices, thus improving the system performance by reducing the load on the central control system.

- **Accuracy:** The accuracy achieved by the digital control systems using smart sensors is much higher than the analogue control systems.
- **Cost-effective:** Though the installation cost of the smart sensor systems is initially higher than that of the conventional systems, the other costs, including the maintenance, addition/replacement of sensors, and the feasibility of the programming, are much better. In addition, the cost is less for long-term operation with the smart sensor systems.
- **Centralized control:** An individual controller can monitor and control many process variables. Although smart sensors have many advantages, a few constraints should be addressed while upgrading the existing smart sensors such as compatibility, discrete wiring, and bus wire failures.

2.3   SMART SENSORS IN IoT APPLICATIONS

IoT has become a significant technology of the information and communication technologies (ICT) field. Most devices in the ICT field, such as smartphones, iPods, tablets, and PCs, are directly connected through the Internet. Besides, a wide range of electronic equipment such as ACs, refrigerators, smart fans, bathroom accessories, and other smart home appliances are also directly associated with the Internet. Likewise, many services are expecting to create huge marketing in the ICT field with IoT potential (Ahlgren et al., 2016). One of the enormous and notable IoT exercises is "Industry 4.0" that mutually created by the German government/Industry/scholarly community. The objective of Industry 4.0 is to associate all the machines in the manufacturing plant through the organization to digitize the entire cycle of processing.

In the typical manufacturing cycle, the structure of the cycle is quite painstakingly planned, but, once it is assembled, it will be fixed for a specific timeframe.

In Industry 4.0, the cyclic process, including the physical arrangement of the plant machine, is changed dynamically (Bledowski et al., 2015). However, the data obtained by the sensors that observe the processes' activities refer to the dynamic operation of the plant. The data from the sensor network not only provide the status of all the hardware in the manufacturing plant but also give the data activities of the workers, the demand for the products, and solicitations from the clients. The production costs are predicted to be reduced remarkably with this fourth industrial revolution.

2.3.1 LIST OF SMART SENSORS USED IN IoT APPLICATIONS

There is a wide range of sensors available in the market as per the requirements and applications. Figure 2.4 describes various smart sensors for various fields of applications and their utilization in respective appliances (Gigli and Koo, 2011).

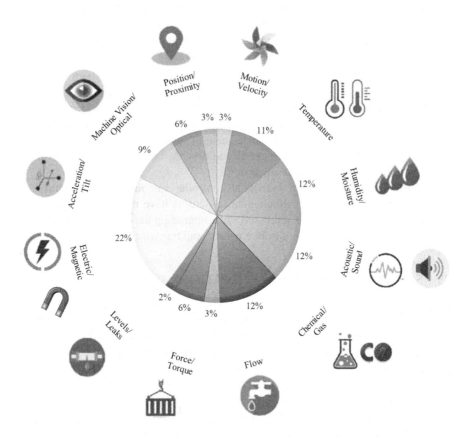

Figure 2.4 List of smart sensors used in IoT applications.

2.3.2   CLASSIFICATION OF SMART SENSORS

Smart sensors can be categorized in many ways that depend upon many criteria. It is a very complex task to classify the sensors. Moreover, all types of sensors can be categorized into two types: passive and active. A passive sensor does not require any external energy. It produces an electrical signal based on an external stimulus. The external stimulus energy is then converted to an output signal by the sensor. Thermocouple, the piezoelectric sensor, is an example of a passive sensor. On the other hand, active sensors require external energy to operate which is called the excitation signal. The active sensors change their properties in response to the external stimulus and produce the output signal. For instance, the thermistor does not generate an electrical signal, but its resistance can be measured by varying the current, voltage across it. Also, sensors can be classified as absolute and relative. A thermistor sensor is an example of an absolute sensor, whereas the thermocouple is an example of a relative sensor. The electrical resistance of the thermistor is directly proportional to the absolute temperature, whereas thermocouple output is proportional to any temperature without referring to a particular baseline. Sometimes sensors are classified based on their properties, specifications, and applications. We have classified the smart sensors into two ways such as based on the characteristics and properties, as shown in Figure 2.5. Sensor classification based on the conversation phenomenon is shown in Figure 2.6. These types of sensors convert the physical properties into chemical or biological forms. Figure 2.7 presents the classification of sensors based on the stimulus.

2.3.3   SENSOR MATERIALS

Nowadays, silicon planar technology has evolved on a large scale, and we are on the verge of disposing of a few sophisticated LSI and VLSI components. Microelectronics can lend to any type of product due to their high precision and performance, and they have been applied for many innovative products worldwide (Popov, 2004). Before the evolution of silicon planar technology, generic non-silicon technologies were used to develop transducers for the sensors. The essential technologies used for sensor materials are listed in this section and depicted in Figure 2.8. These are as follows:

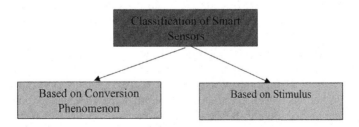

Figure 2.5   Classification of sensors (based on the characteristics and properties).

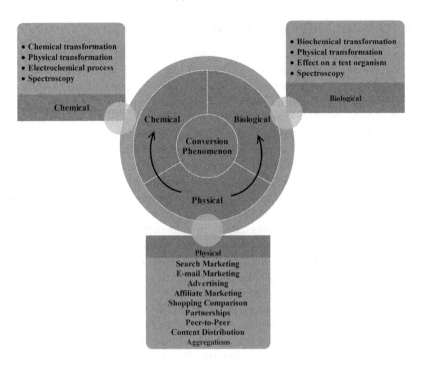

Figure 2.6   Classification of smart sensors based on conversion phenomenon.

- Piezoelectric material
- Polymers
- Metal oxides
- Semiconductors
- Thick- and thin-film materials
- Optical glass fibers.

2.4   ROLE OF SENSOR NETWORKS IN VARIOUS APPLICATIONS

Wireless sensor networks (WSNs) consist of sensor nodes that can communicate to collect the data wirelessly from the environment deployed for a specific application. WSNs are self-configured to monitor real-time physical conditions like six domain signals, as discussed in Section 2.2. The collected data are passed to the primary location or base station, where it can be observed, analyzed, and processed to take an intended action. Though the WSNs are widely used in many applications, there are certain limitations to be focused on such as limited memory, battery power, communication range, location of the application, computing methods, bandwidth, and privacy. Owing to these limitations, WSNs require nonconventional design methods to enable customized applications (Boubiche et al., 2018). There is a wide range of

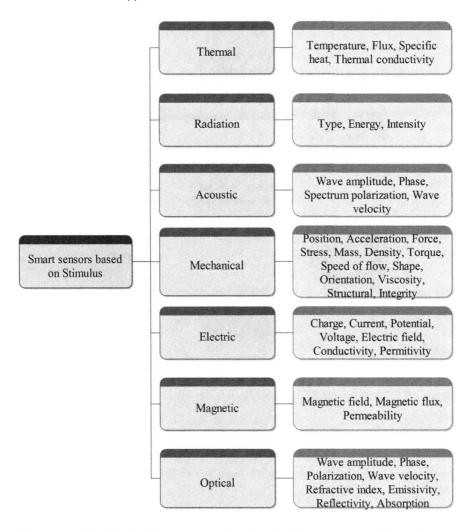

Figure 2.7 Classification of smart sensors based on stimulus.

applications including military, satellites and spaceships, automobile and transportation, health applications, home automation, structural monitoring, environment conditioning, and agriculture. A **smart home** is an example of the momentum that IoT has brought to the technological world. A smart home is a residence connected with smart sensors, devices, systems, equipment, appliances that can be accessed, monitored, and controlled remotely using the Internet. In a survey conducted by Statista in August 2020 (as described in Figure 2.9), the value of the global smart home market in 2017 is around $46.16 billion and is expected to reach about $92.74 billion in 2020. Since security and privacy are the key features of the smart home system, many householders are keen to purchase this system. Also, the report shown in Figure 2.9

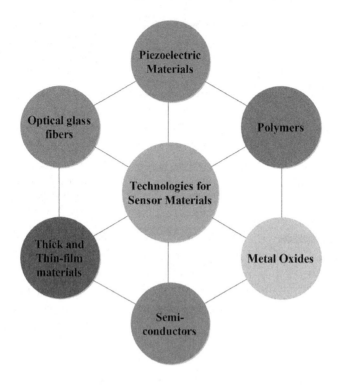

Figure 2.8   The essential technologies for smart sensor materials.

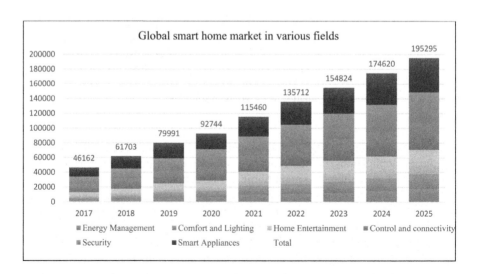

Figure 2.9   Global smart home market in various fields.

indicates that the adoption of smart home systems helps in energy-saving and, thus, the environment. As a result, it leads large commercial companies such as Google, Apple, and Amazon into the smart home market with their software, hardware, and services. They have introduced voice-controlled devices into the market with a wide range of smart sensor technologies that bring many advantages to householders and society. The traditional home usually contains the apparatus and appliances that can be operated manually and locally with pushbuttons, switches, knobs, and so on. Simultaneously, a smart home incorporates Internet-connected appliances and apparatus that can be interacted intelligently and operated remotely. Hence, a smart home can be defined as a living place, house, apartment, and so on, which involves remote monitoring, controlling, and accessing the devices with smart sensors, and data communication technologies using the Internet (Kodali and Soratkal, 2016; Liu et al., 2019).

Smart sensors are the critical components of the smart home system in sensing, measuring, and detecting the changes from various environment domains (Pirbhulal et al., 2017). The data transmitted to the actuators can take the physical actions such as switching ON/OFF the lights, increasing/decreasing the speed of the fans, raising the safety alarms, activating/deactivating the heating appliances, and controlling the speed of the motors, volume of the speakers, and so on. Then, the data obtained or collected from the smart sensors must be compiled and transmitted to connect the network and the environment. Such operations can be performed by gateways, which are responsible for interfacing with the environment for transmitting compiled data in and out to the Internet and to enable services (Rusitschka et al., 2010). Gateways could be a smartphone, a computing device like Arduino, Raspberry Pi, or any dedicating device (Bera et al., 2014; Carvalho et al., 2017). These gateways need to be connected to the Internet through a wired (ethernet) or wireless network (Wi-Fi).

Many of the gateways perform the following functions when they receive signals from the connected devices:

- Notifications
  - Typically, a text message/alert message pop-up is displayed through a smartphone, like the flooding of water from the tank.
- Automation
  - Predefined conditions can be set to perform a specific action, like turning ON lights when detects a motion.
- Local control
  - In the absence of the Internet or cloud service(s), the interface can be done through a desktop or mobile application.
- Cloud service
  - Many of the gateways utilize the cloud service(s) for device management, data storage, user interfacing, and data communication.
- API/SDK
  - A customized functionality can be built with the application programming interface (API), and/or software development kit (SDK) is used.

2.4.1 SMART HOME APPLICATION – A CASE STUDY (OUR DEVELOPED MODEL)

Recently, we have developed a smart home control system. The developed model is shown in Figure 2.10. It consists of a smartphone with APK installed, which can be used to turn ON/OFF the lights, increase/decrease the illumination of the lights, and increase/decrease the speed of the induction motor, ceiling fan, and water pump. To interface the appliances with the controller, Google Firebase is used as Cloud Service. The controlled signals can be provided to the devices through Arduino UNO to an open-source Lua-based firmware and development board NodeMCU, which operate with the ESP8266 Wi-Fi SoC. An API is developed using MIT application inventor to control the connected appliances through the interface built with the Google Firebase. The built-in home automation system consists of NodeMCU with ESP2866 Wi-Fi board, which can be powered through a USB port. It can be interfaced with different home appliances such as a bulb, motor, ceiling fan, pump, and tube light.

These appliances can be controlled through the relay channels connected with a 230V AC supply, and the controlled signals can be given through a smartphone installed with an API. The control signals can be sent to various appliances by setting up the number of general purpose input output (GPIO) pins of ESP2866 which acts as a webserver for the system. Once the GPIO pins are configured, a hypertext transfer protocol (HTTP) server can be created to change the status of the GPIO and thereby can control the appliances connected. Authentication can be provided for the control system through service set identifier (SSID) and password inside the program that must be debugged with NodeMCU so that it can attempt to discover an access point once it is powered up. The complete system can be configured either through a wired or through a wireless medium.

The credentials required for authentication control operation are provided in the program with the necessary library functions and restored in an Arduino-UNO board, which further is interfaced with ESP2866. Then, the access point connects, only when the correct credentials are given, and then it becomes a part of the WLAN. It

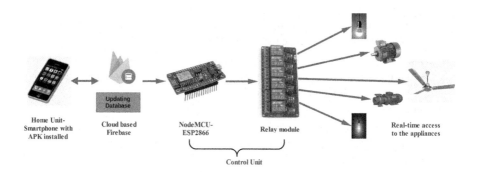

Figure 2.10   Control process of the appliances from the home unit to the control unit.

requires to have a local IP address for the server, which is set by a portal address of 80, and that can be displayed in the serial monitor with the Arduino board. The data sent by the client are received using this port address 80. The command signals can be sent to modify the status of the configured GPIO pins to control the appliances through the relay modules. Here, the relay acts as a switching device to control the home appliances that function at the voltage level of 230V AC. The API required to send the command signals through a smartphone is developed using an open-source web application MIT App inventor as described in Figure 2.10. Since it is a cloud-hosted NoSQL database that manages SDKs with the users and allows building the required applications without the servers, when a user is offline, it is possible to use SDKs as a local server and can store the changes made by using the local cache. However, when the device is online, automatically the data will be synchronized with an intuitive authentication process.

As this control system is developed using NodeMCU (ESP2866) to control the home appliances using an android application, the command signals can be sent to the firebase cloud service with the given URL with an authorization token. At that instant, the NodeMCU is connected to the Internet and starts receiving the command signals from the firebase with the given URL and authorization token. Since the database uses a real-time database, the controller sends the commands to MCP23017, which is a port-extender. It is an I2C communication device that uses both SDA and SCL pins for communication with the controller. Based on this given command, the switching ON/OFF action of lights/fans is controlled as demonstrated in Figure 2.11. Then, the commands are given to the relay module to the corresponding light/fan circuit. The fan speed regulation is obtained based on the slider position set in the android application. Different levels of speed are obtained by changing the slider positions. By doing so, we could achieve four levels of speeds for the ceiling fans. Also, the illumination of the lamps could vary with the developed controller. The images of controlled devices are shown in Figure 2.12. The PCB-based circuit of the proposed model is shown in Figure 2.13, which is successfully installed in one of our laboratories to control six ceiling fans and six tube lights.

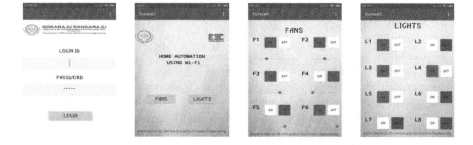

Figure 2.11   The developed app for user interface using a smartphone.

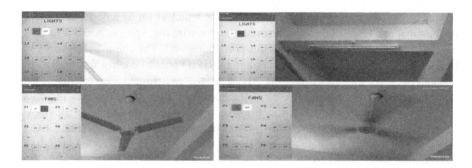

Figure 2.12   Images of controlled device using smartphone (IoT).

Figure 2.13   The PCB-based model of the developed smart home system controller.

2.4.2   CHALLENGES AND ISSUES IN DEVELOPING SMART HOMES

Many challenges and issues need to be addressed on a prioritized basis in developing a smart home system. The following issues are the key issues to be attended before designing a smart home system model:

- Interoperability
- Security and privacy
- Reliability
- Usability.

Interoperability: The devices, smart sensors, and components involved in the smart home system must be adequately integrated to work effectively. The ability of all these devices to work together is called Interoperability. As discussed in the case study, a NodeMCU is programmed to receive and send the command signals to control the appliances connected through API configured in the smartphone and further authenticated through the Firebase. In an instant of operation, if any of the devices and/or services are not responding to the command signals, the desired application cannot be executed. Therefore, Interoperability is an important and necessary issue in evolving a smart home from a normal home in a more intelligent way where all the devices and services collaborate in achieving a unified task.

Security and Privacy: The IoT consists of many heterogeneous devices in the smart home system, through which the data can be gathered, exchanged, and stored in the environment. The data stored could be personal data such as photos, videos, personal information, and authenticated credentials, which are meant to serve in private spaces that lead to security risks (Sliwa and Benoist, 2011). These risks may vary from system to system and service to service, the protocols that devices follow, and the individual behavior of the components. Since the data are collected with many devices, they may open the entry points for malicious software from different threat sources. The risk factor may vary from low to critical, depending upon the behavior of the heterogeneous device. Therefore, it is essential to have a centralized network that can access the privacy and risk factors to develop an effective risk management strategy.

Reliability: The smart home system performance depends upon the accurate functioning of the IoT devices, which should not either be failed or malfunctioned. In comparison to conventional devices, IoT devices are complex and less reliable in certain conditions (Sikder et al., 2018). For instance, in a condition of water overflows in a tank, when the motor is supposed to turn OFF automatically with the sensed data, the failure of the sensor may lead to the malfunction of the motor operation. Though this may be accepted in the energy domain, such malfunctioning is not acceptable in the health-care domain (Kan et al., 2015; Webster and Eren, 2015). Hence, reliability is a significant concern in the smart home system and other comprehensive applications.

Usability: The smart home system provides a user-friendly atmosphere to the customers (or householders) to use the appliances. Since the IoT devices can be used by experts and non-experts, it is vital how an end user would interact with the smart devices and use the IoT services. The ease of use and simplicity can be defined as usability in perspective to the end user. The usability should be provided to the end user without compromising the security and risks, which is one of the significant challenges of the developer (Daidone et al., 2011).

2.5 CONCLUSION

The application range of smart sensors is increasing day by day with the advanced technology developed by the manufacturers. For many years, smart sensors have been manufactured with the use of sensors and MCU only. Nevertheless, the increasing demand for smart applications allowed many big companies to invest in the smart sensor market with advanced technologies. Recent trends in MEMS technology have brought many advantages, such as improved accuracy and resolution, increased operating range, ease of integration of the devices, nanostructures, and cost reduction through the improved processors. With complex and innovative sensor technologies, it is possible to enable more new parameters. The performance of the smart sensors can be improved with the integration of the MCUs, DSPs, ASICs, and FPGAs, which further simplifies the sensor interface for the specific application and allows faster operation. The application of smart sensors in the IoT frontier allows an easy way of controlling appliances and devices. However, there are a few network issues and challenges in adopting smart sensors in IoT devices, which can be minimized by proper risk and privacy management system. The combination of the MEMS technology and wireless technology can lead to a new era of smart sensors.

REFERENCES

Ahlgren, B., Hidell, M., & Ngai, E. C. H. (2016). Internet of things for smart cities: Interoperability and open data. *IEEE Internet Computing*, 20(6), 52–56.

Bera, S., Misra, S., & Rodrigues, J. J. (2014). Cloud computing applications for smart grid: A survey. *IEEE Transactions on Parallel and Distributed Systems*, 26(5), 1477–1494.

Bledowski, K. The Internet of Things: Industrie 4.0 vs. the Industrial Internet, July 2015. URL https://www. mapi.net/forecasts-data/internet-things-industrie-40-vs-industrial-internet.

Boubiche, D. E., Pathan, A. S. K., Lloret, J., Zhou, H., Hong, S., Amin, S. O., & Feki, M. A. (2018). Advanced industrial wireless sensor networks and intelligent IoT. *IEEE Communications Magazine*, 56(2), 14–15.

Carvalho, G. H., Woungang, I., Anpalagan, A., Jaseemuddin, M., & Hossain, E. (2017). Intercloud and HetNet for mobile cloud computing in 5G systems: Design issues, challenges, and optimization. *IEEE Network*, 31(3), 80–89.

Colakovi¢. A and Hadziali¢. M (2018). Internet of Things (IoT): A review of enabling technologies, challenges, and open research issues, *Computer Networks*, 144, 17–39.

Daidone, R., Dini, G., & Tiloca, M. (2011, June). On experimentally evaluating the impact of security on IEEE 802.15. 4 networks. In *2011 International Conference on Distributed Computing in Sensor Systems and Workshops (DCOSS)* (pp. 1–6). IEEE.

Gad-el-Hak, M. (Ed.). (2005). *MEMS: Introduction and Fundamentals*. CRC Press, Boca Raton, FL.

Geirinhas, R. H. (2007). A Contribution to the IEEE STD. 1451.2–1997 Revision and Update. AFRICON 2007, Windhoek, 2007, pp. 1–7.

Gigli, M., & Koo, S. G. (2011). Internet of things: Services and applications categorization. *Advanced Internet of Things*, 1(2), 27–31.

Gubbi, J., Buyya, R., Marusic, S., & Palaniswami, M. (2013). Internet of Things (IoT): A vision, architectural elements, and future directions. *Future Generation Computer Systems*, 29(7), 1645–1660.

Khan. J.Y and Yuce, M. R. (2019). *Internet of Things (IoT): Systems and Applications*, 1st ed. Jenny Stanford, Singapore.

Kan, C., Chen, Y., Leonelli, F., & Yang, H. (2015, August). Mobile sensing and network analytics for realizing smart automated systems towards health internet of things. In *2015 IEEE International Conference on Automation Science and Engineering (CASE)* (pp. 1072–1077). IEEE.

Kodali, R. K., & Soratkal, S. (2016, December). MQTT based home automation system using ESP8266. In *2016 IEEE Region 10 Humanitarian Technology Conference (R10-HTC)* (pp. 1–5). IEEE.

Liu, X., Lam, K. H., Zhu, K., Zheng, C., Li, X., Du, Y., ... & Pong, P. W. (2019). Overview of Spintronic Sensors With Internet of Things for Smart Living. *IEEE Transactions on Magnetics*, 55(11), 1–22.

Liu, X., & Baiocchi, O. (2016, October). A comparison of the definitions for smart sensors, smart objects and Things in IoT. In *2016 IEEE 7th Annual Information Technology, Electronics and Mobile Communication Conference (IEMCON)* (pp. 1–4). IEEE.

Pirbhulal, S., Zhang, H., E Alahi, M. E., Ghayvat, H., Mukhopadhyay, S. C., Zhang, Y. T., & Wu, W. (2017). A novel secure IoT-based smart home automation system using a wireless sensor network. *Sensors*, 17(1), 69.

Popov, V. N. (2004). Carbon nanotubes: Properties and application. *Materials Science and Engineering: R: Reports*, 43(3), 61–102.

Padmalaya Nayak, (2016) "Internet of Things Issues, Challenges and Applications" Handbook of Research on advanced WSN Applications, Protocols and Architecture, IGI Global ISSN-2327-3305, pp. 353–368.

Rusitschka, S., Eger, K., & Gerdes, C. (2010). Smart grid data cloud: A model for utilizing cloud computing in the smart grid domain. In *2010 First IEEE International Conference on Smart Grid Communications* (pp. 483–488). IEEE.

Safak, M. (2014). Wireless sensor and communication nodes with energy harvesting. *Journal of Communication, Navigation, Sensing and Services (CONASENSE)*, 1(1), 47–66.

Sikder, A. K., Petracca, G., Aksu, H., Jaeger, T., & Uluagac, A. S. (2018). A survey on sensor-based threats to internet-of-things (IoT) devices and applications. arXiv preprint arXiv:1802.02041.

Sliwa, J., & Benoist, E. (2011). Pervasive computing–the next technical revolution. Developments in E-systems Engineering.

Webster, J. G., & Eren, H. (2015). *The E-Medicine, E-Health, M-Health, Telemedicine, and Telehealth Handbook*. Volume II: Telehealth and Mobile Health, CRC Press, Boca Raton, FL.

3 Stochastic Modeling in the Internet of Things

Srinivas R. Chakravarthy
Kettering University
Flint, Michigan, USA

CONTENTS

3.1 Introduction 35
3.2 Model Assumptions...38
3.3 Steady-State Analysis for One-Service Provider Case....................................40
    3.3.1 Steady-State Equations 40
    3.3.2 System Measures for One-Service Provider Case 41
3.4 Steady-State Analysis for Two-Service Provider Case 42
    3.4.1 Steady-State Equations for the Two-Service Provider Case..............44
    3.4.2 System Measures for Two-Service Provider Case...............................46
3.5 Systems with More Than Two-Service Providers 47
3.6 Illustrative Examples 47
3.7 Concluding Remarks 60
References..61

3.1 INTRODUCTION

The current technology has permeated everything we do in our day-to-day activities. Since the introduction of the term "Internet of Things" (*IoT*) by Ashton (2009), this concept has become a hot topic in almost all fields. According to a recent report, there were 7 billion *IoT* devices connected to the Internet in 2018, and in just 2 years, it is expected to exceed 20 billion devices according to Mahmood (2020) and 50 billion devices according to Rathod & Chowdhary (2019). These devices are used in a wide variety of applications such as automotive, home appliances, medical devices, and many more. The beauty of *IoT* lies not in its control alone but rather in its intelligence. A large number of data are collected through *IoT* and its intelligence, inventory of common household items can be periodically checked, and where needed automatic replenishment of items falling below a certain point can be done even without the knowledge of the household members.

IoT can be thought of as a network of many physical elements such as sensors (to collect vital information), identifiers (to identify the sources of data), software (needed in data analysis), and Internet connectivity (for communication purposes). This network of elements enable *IoT* to perform its intended tasks seamlessly. As more and more people use Internet for business, recreation, education, health care, and other purposes, it is not surprising to see the exponential explosion in the number of devices being used.

The success of *IoT* depends on a number of factors such as the architecture for the *IoT* platform, reliable software, efficient algorithm, state-of-the-art communication technologies, minimal human intervention, security, integration of hardware and software from various vendors, and remote device management, among other things. These and other key concepts can be found in the literature (see, e.g., Balas, Solanki, Kumar, & Ahad (2020); Balas, Solanki, Kumar, & Srivastava (2020); Evans (2011); Firouzi, Chakrabarty, & Nassif (2020); Hassija, Chamola, Saxena, Jain, Goyal & B. Sikdar (2019); Kanagachidambaresan, Anand, Balasubramanian, & Mahima (2020); Mahmood (2019, 2020); Milenkovic (2020); Mohamed (2019); Raj, Chatterjee, Kumar, & Balamurugan (2020); Rayes & Salam (2019).

Opening up households and businesses to communicate with other *IoT* devices poses other problems including security. The requirements for the successful operation of *IoT* and challenges today are well established in the literature (see, e.g., Mohamed (2019)).

Sensing-as-a-Service, known as *SEaaS*, can be viewed from many angles. One such angle is to generate batches of demands from sources like households and small businesses.

It should be pointed out that the purpose of this chapter is not to discuss various layers, such as physical (mostly consisting of senors needed to gather data for which this layer is used), communication (considered as the backbone for *IoT* has hardware such as Wi-Fi, 5G, RFID, etc., to communicate the data to other devices) and application (providing network services to the end host's applications). The details on these, for example, can be found in the references mentioned earlier. Rather, the purpose of this chapter is to introduce Neuts' versatile Markovian point process as a way to model the batches of demands occurring from various *IoT* devices in the context of *SEaaS*. This is especially important now as more and more people including the generations of the Boomers (ages 55–73) and the Silent (ages 74–91) are using some form of *IoT* devices to help them manage their daily routine. Having stochastic models to better understand the demands and meeting them under random environment will help the service providers (*SPs*)to plan their resources accordingly.

While *IoT* has been studied in many perspectives such as cloud layer, fog layer, and device layer, the key thing in *IoT* is to apply analytic to big data to get insights and take appropriate actions promptly. One of the main purposes of *IoT* analytic is to create and sustain smarter environments such as smarter homes, smarter hospitals, smarter service sectors, and so on to provide timely and quality services to people. Toward this end, one has to develop suitable queueing models.

There is an extensive literature on queueing models (see, e.g., Alfa (2010); Bhat (2015); Dudin, Klimenok, & Vishnevsky (2020); Harchol-Balter (2013); He (2014)), and only recently, there appears to be an increasing interest in applying queueing to

IoT in the context of (i) application flows in broadband networks (see, e.g., Kua, Nguyen, Armitage, & Branch (2017)); (ii) relay technology (see, e.g., Li, Zaho, Yu, & Huang (2016); (iii) emergency packet transmissions (e.g., Qiao, Qiu, Han, Ma, & Huang (2017); (iv) fog-edge based middleware distributed service provisioning (see, e.g., Rathod & Chowdhary (2019); (v) priority based push-out management scheme (see, e.g., Salameh, Awad, & AbuAlrub (2019); (vi) delay sensitive traffic (see, e.g., Sharma, Kumar, Gowda, & Srinivas (2015); (vii) health-care system (see, e.g., Strielkina, Uzun, & Kharchenko (2017); (viii) automatic queue monitoring system in a store Viriyavisuthisakul, Sanguansat, Toriumi, Hayashi, & Yamasaki (2017); (ix) providing an assessment of the data service Volochiy, Yakovyna, & Mulyak (2017); and (x) improving the quality of service where fog computing is employed Yousefpour, Ishigaki, Gour, & Jue (2018).

However, to the best of our knowledge, this is the first attempt to incorporate queueing modeling in *IoT* from the point of view of providing services (to customers) by properly allocating the requisite resources and to incorporate correlation in the inter-demand (or inter-arrival) times. Further, to our knowledge, there is not much literature in *IoT* with regard to the study of queueing models useful in providing smarter services. For example, one of the main purposes of having smarter devices at home is to automatically procure items such as refills on medication, consumable items such as milk, bread, cereals, pet foods, and other commonly seen items in the house. By having such smarter devices, one need not worry about missing these items to be procured regularly.

At the lowest level of the hierarchy, each household/commercial establishment (also known as customers) will have a client resource unit (*CRU*). The main purpose of the *CRU* is to consolidate the need and route the orders to appropriate *SP*.

The goals of this study are as follows. First, the batch Markovian arrival process (*BMAP*), a versatile Markovian point process introduced by Neuts (1979), is used in the context of *SPs* for the *IoT*. Second, the minimum number of resources needed to achieve a given level of throughput in a finite capacity *SP* system is obtained. This optimum number of *SPs* is used when dealing with more than one *SP* system to make proper comparisons among various scenarios.

In stochastic modeling, matrices play an important role. Hence, it is imperative to introduce some key notations and definitions, in order to be used in the rest of the chapter:

- *e* is a column vector of 1's. The dimension of this and other vectors/matrices defined below will be clear from the context. However, where more clarity is needed, the dimension will be displayed. As an example $e(m)$ will mean *e* of dimension m.
- *I* is an identity matrix.
- $\Delta(E_0, E_1, \ldots, E_n)$ stands for a diagonal matrix with (block) elements, $E_0, E_1, \ldots, E_n$, which themselves are matrices.
- The symbols, $\otimes$ and $\oplus$, respectively, define the Kronecker product and sum of matrices. A few key references on these can be found in Graham (1981); Marc & Minc (1964); Steeb & Hardy (2011).

This chapter is organized as follows. In Section 3.2, the assumptions of the model under study are described. The one-*SP* case is studied in Section 3.3, and in Section 3.4, the two-*SP* case is discussed. Owing to the exponential growth of the state space, simulation-based analysis of the models dealing with more than two *SPs* is spelled out in Section 3.5. Illustrative numerical examples for all the models presented are discussed in Section 3.6. Some concluding remarks including future work are given in Section 3.7.

3.2   MODEL ASSUMPTIONS

We assume that there is one *CRU* which is responsible to send the demands for processing to one (and only one) of the *SP* at its disposal. Demands arrive at the *CRU* according to a *BMAP* with parameter matrices $\{D_k\}_{k=0}^{\infty}$ such the matrix $D = \sum_{k=0}^{\infty} D_k$ governs the underlying Markov chain of the *BMAP*. While D_0 governs transitions corresponding to no arrivals, the matrix D_k, $k \geq 1$, governs transitions corresponding to arrival of batch of size k. For all practical purposes, the batch size will not exceed a finite number, and hence, it is assumed that the maximum size is $r < \infty$. That is, $D_k = 0, k > r$.

BMAP is a versatile point process introduced initially by Neuts with complex notation. However, the notation was simplified more than a decade later by Neuts and his research associates in Lucantoni, Meier-Hellstern, & Neuts (1990). Since then, this description of *BMAP* is the standard by which such processes are presented and studied. The beauty of a *BMAP* is to incorporate correlation that is naturally inherent in the inter-arrival times in most applications, especially in, *IoT* as the demands (or orders) arrive from various sources. Even if these sources generate the orders independently (which again may not be the case in practice), the cumulative orders arriving at the *CRU* may not be generated as independent and identically distributed processes (unless each source is modeled using a Poisson process). Thus, the need to use *BMAP* in *IoT* context is very critical and important.

To get an expression for the (batch) arrival rate in the *BMAP* under consideration, the steady-state probability vector of the irreducible generator D that governs the *BMAP* is needed. Let π be that vector such that

$$\pi D = 0, \quad \pi e = 1. \tag{3.1}$$

The quantity $\lambda_g = \pi \sum_{k=1}^{\infty} D_k e$ gives the arrival rate of batches and $\lambda = \pi \sum_{k=1}^{\infty} k D_k e$ is the expected number of customer arrivals per unit time.

Suppose that X denotes the time to the first (group) arrival in the *BMAP* process. By taking the initial probability vector of the *BMAP* process to be $\beta = \frac{1}{\lambda_g} \pi(-D_0)$ (see equation (3.1) for π), it is easy to verify (see, e.g., Chakravarthy (2001, 2010)) that the probability density function of X is given by

$$f(x) = \beta e^{D_0 x}(-D_0)e, \ x \geq 0, \tag{3.2}$$

and the joint probability density of function of the first two intervals between (group) arrivals is given by

$$f(x_1,x_2) = \beta e^{D_0 x_1}(D - D_0)e^{D_0 x_2}(-D_0)e, \; x_1 \geq 0, \; x_2 \geq 0. \tag{3.3}$$

The mean μ_X, the variance (σ_X^2), and one lag correlation ($\hat{\rho}$), are obtained as (see, e.g., Chakravarthy (2001, 2010))

$$\mu_X = \frac{1}{\lambda_g}, \; \sigma_X^2 = \frac{2}{\lambda_g}\pi(-D_0)^{-1}e - \frac{1}{\lambda_g^2}, \tag{3.4}$$

and

$$\hat{\rho} = \frac{\lambda_g \pi[(-D_0)^{-1}(D - D_0)](-D_0)^{-1}e - 1}{2\lambda_g \pi(-D_0)^{-1}e - 1}. \tag{3.5}$$

BMAP, which includes *MAP* as a special case, has been used extensively in stochastic modeling, specifically in telecommunications, inventory, reliability, and queueing. The reader is referred to Alfa (2010); Artalejo, Gomez-Correl, &He (2010); Chakravarthy (2001, 2010, 2015); Lucantoni et al. (1990); Lucantoni (1991); Neuts (1979, 1989, 1992, 1995) for details on *MAP* and *BMAP* and other key references.

There are n *SPs* each with a finite number of resources. A demand of size k requires exactly k resources for processing. Assume that SP_i, $1 \leq i \leq n$, has N_i resources. Thus, the ith *SP*, SP_i, can have at most N_i items being processed at any given time. Each *SP* will process the items within a batch as soon as they are received by allocating the requisite number of resources for the batch (either fully or partially, depending on the availability of the resources). Thus, there is no queue either at the *CRU* or at the *SP* area, and hence there is a possibility of a loss of one or more items within a batch due to lack of resources.

An arriving demand is assigned to only one *SP* as pointed out earlier. The assignment of a *SP* is done as follows. Starting with SP_1 and ending with SP_N, the first available *SP* that can handle the entire batch of an arriving demand will be used. If no *SP* has enough resources to handle an arriving demand, the *CRU* will accept the demand to the level at which it can be served by the first *SP* starting from 1 to n and the rest of the items in the demand will be considered lost. For example, if $n = 3$ and if the number of resources available in these three *SPs* are, respectively, 2, 4, and 5. If an arriving demand is of size 4, the entire demand will be sent to SP_2. However, if the arriving demand size is 7, none of the three can handle the entire batch, and hence 2 out of 7 will be allotted to SP_1 and the rest will be lost. This is mainly due to our assumptions of choosing the *SP* in that specific order. These assumptions along with the one wherein a batch cannot be split to be sent to more than one service can all be relaxed and still adopt the same methodology to analyze such models. Some of the details are provided in Section 3.7.

We assume that each resource takes an exponential amount of time to process an item. The parameter of the exponential distribution depends on the *SP*, and thus the average processing rate of an item (within a batch) in SP_i is taken to be

μ_i, $1 \le i \le n$, per unit of time. Thus, a batch of k items will consume k resources of the *SP* where the batch is assigned. Note that the model studied in this chapter is always stable.

In this chapter, it is assumed that as soon as an item is processed, it will be sent out as opposed to waiting for the rest of the items, if any, in that batch to be finished. If the entire batch needs to be sent out as a whole, then one needs to incorporate the resequencing concept and will be addressed elsewhere. Typically, in practice, one sees items arrive at the customer's place at different times even if all the items within a batch are ordered at the same time, and hence this policy is adopted here.

3.3 STEADY-STATE ANALYSIS FOR ONE-SERVICE PROVIDER CASE

In this section, the model corresponding to the one-*SP* system is discussed. Since there is only one *SP*, the maximum number of resources is given by N. For this case, the state space of the Markov process governing the system

$$\Omega_1 = \{(i,k) : 1 \le k \le m, 0 \le i \le N\}.$$

We denote by $\underline{i}$, $0 \le i \le N$, the set of states corresponding to exactly i items that are being processed by SP_1 with the *BMAP* process in various phases. Define

$$
A_0^{(1)} =
\begin{pmatrix}
D_0 & D_1 & D_2 & \cdots & D_r & & & & & & \\
\mu_1 I & D_0 & D_1 & D_2 & \cdots & D_r & & & & & \\
2\mu_1 I & D_0 & D_1 & D_2 & \cdots & & D_r & & & & \\
& \ddots & \ddots & \ddots & \ddots & \ddots & & & & & \\
& & & & & \ddots & (N-r)\mu_1 I & D_0 & \cdots & D_{r-1} & \tilde{D}_r \\
& & & & & & & \ddots & & \ddots & \\
& & & & & & & & & D_0 & \tilde{D}_1 \\
& & & & & & & & & N\mu_1 I & D
\end{pmatrix},
$$

$$\tag{3.6}$$

where $\tilde{D}_k$ is given by

$$\tilde{D}_k = \sum_{j=k}^{r} D_j, \quad 1 \le k \le r. \tag{3.7}$$

The generator, say, $Q^{(1)}$, of the Markov process governing the system with only one-*SP* can be verified to be

$$Q^{(1)} = A_0^{(1)} - \Delta(0, \mu_1 I, 2\mu_1 I, \ldots, N\mu_1 I). \tag{3.8}$$

3.3.1 STEADY-STATE EQUATIONS

Let $x = (x_0, x_1, \ldots, x_N)$ denote the steady-state probability vector of $Q^{(1)}$. That is, the vector, x, of dimension $(N+1)m$ satisfies

$$xQ^{(1)} = 0, \quad xe = 1. \tag{3.9}$$

In terms of vectors and matrices of dimension m suitable for numerical implementation, the equation given in (3.9) can be rewritten as

$$x_0 D_0 + \mu x_1 = 0,$$

$$x_j(D_0 - j\mu I) + \sum_{k=1}^{\min\{j,r\}} x_{j-k} D_k + (j+1)\mu x_{j+1} = 0, \ 1 \leq j \leq N-r,$$

$$x_j(D_0 - j\mu I) + \sum_{k=1}^{r} x_{j-k} D_k + (j+1)\mu x_{j+1} = 0, \ N-r+1 \leq j \leq N-1,$$

$$x_N(D - N\mu I) + \sum_{k=1}^{r} x_{N-k} \tilde{D}_k = 0,$$

$$\sum_{k=0}^{N} x_k e = 1.$$

(3.10)

The equations given in (3.10) can be easily implementable for the numerical evaluation of the steady-state probability vector. Depending on whether one is willing to store many inverses as opposed to only a few. For example, by computing and storing only three inverses: $(-D_0)^{-1}$, $(N\mu I - D_0)^{-1}$, and $(N\mu I - D)^{-1}$, one can implement (block) Gauss-Seidel iterative procedure to compute the vector x as follows:

$$x_0 = \mu x_1 (-D_0)^{-1},$$

$$x_j = \left[(N-j)x_j + \sum_{k=1}^{\min\{j,r\}} x_{j-k} D_k + (j+1)\mu x_{j+1} \right]$$
$$\times (N\mu I - D_0)^{-1}, \quad 1 \leq j \leq N-r,$$

$$x_j = \left[(N-j)\mu x_j + \sum_{k=1}^{r} x_{j-k} D_k + (j+1)\mu x_{j+1} \right]$$
$$\times (N\mu I - D_0)^{-1}, \quad N-r+1 \leq j \leq N-1,$$

$$x_N = \left[\sum_{k=1}^{r} x_{N-k} \tilde{D}_k \right] (N\mu I - D)^{-1},$$

(3.11)

which is subject to the normalizing condition.

It is important in any numerical computation to have some internal accuracy checks. In the current case, it is easy to verify the following intuitively obvious result, which can be used as a key accuracy check:

$$\sum_{j=0}^{N} x_j = \pi,$$

(3.12)

where π is as given in (3.1).

3.3.2   SYSTEM MEASURES FOR ONE-SERVICE PROVIDER CASE

In this section, a few performance measures to qualitatively describe and compare various models studied in this chapter are listed.

(1) The idle probability, P_{idle}, of the system is given by

$$P_{idle} = x_0 e$$

(2) The mean number in the system (which is also the mean number of busy resources), μ_{NS}, is calculated as

$$\mu_{NS} = \sum_{i=1}^{N} i x_i e.$$

(3) The variance of the number in the system (which is also the mean number of busy resources), σ_{NS}^2, is calculated as

$$\sigma_{NS}^2 = \left(\sum_{i=1}^{N} i^2 x_i e \right) - \mu_{NS}^2.$$

(4) The coefficient of variation of the number busy resources is obtained as $\frac{\sigma_{NS}}{\mu_{NS}}$.

(5) The probability, P_{loss}, that a loss occurs at an arrival epoch, is obtained as

$$P_{loss} = \frac{1}{\lambda_g} \sum_{k=1}^{r} \sum_{j=k}^{r} x_{N-r+j} D_{r-j+k} e.$$

(6) The mean number of lost items, μ_{loss}, at an arrival epoch, is calculated as

$$\mu_{loss} = \frac{1}{\lambda_g} \sum_{k=1}^{r} k \sum_{j=k}^{r} x_{N-r+j} D_{r-j+k} e.$$

(7) The throughput, γ, which is the rate at which the items are processed per unit of time, is computed as

$$\gamma = \lambda - \lambda_g \mu_{loss}.$$

Other measures such as the percentiles and the mode for the number of busy resources can be obtained with the standard formulas. The details are omitted.

3.4 STEADY-STATE ANALYSIS FOR TWO-SERVICE PROVIDER CASE

Here, the model corresponding to the two-SP system is discussed. For the sake of simplicity in presenting the key points, it is assumed that both the SPs have the same amount of resources. The relaxation of this assumption and its impact are discussed later. That is, here it is assumed that $N_1 = N_2 = N$. It is easy to see that the state space of the Markov process governing the system can be written as

$$\Omega_2 = \{(i, j, k) : 1 \leq k \leq m, 0 \leq i, j \leq N\}.$$

Denote by $\underline{i} = (i, j, k)$, $0 \leq j \leq N$, $1 \leq k \leq m$, for $0 \leq i \leq N$ the set of states of dimension $(N+1)m$, corresponding to the case where exactly i items are being processed by SP_2, j items are processed by SP_1, and the demand process is in various states.

For use in sequel, the needed auxiliary matrices to obtain the generator of the Markov process governing the two-SP model are listed:

$$
A_0^{(2)} =
\begin{array}{c}
\\ 0 \\ 1 \\ 2 \\ \vdots \\ N-r \\ \vdots \\ N-1 \\ N
\end{array}
\begin{pmatrix}
D_0 & D_1 & D_2 & \cdots & D_r & & & & & & & \\
\mu_1 I & D_0 & D_1 & D_2 & \cdots & D_r & & & & & & \\
& 2\mu_1 I & D_0 & D_1 & D_2 & \cdots & D_r & & & & & \\
& & \ddots & \ddots & \ddots & \ddots & \ddots & & & & & \\
& & & & & & \ddots & (N-r)\mu_1 I & D_0 & \cdots & D_{r-1} & D_r \\
& & & & & & & & \ddots & \ddots & & \\
& & & & & & & & & & D_0 & D_1 \\
& & & & & & & & & & N\mu_1 I & D_0
\end{pmatrix}
$$

(3.13)

The matrix $\tilde{A}_k^{(2)}$, $1 \leq k \leq r$, is obtained from $A_0^{(2)}$ by replacing the last non-zero (block) entries with the following non-zero (block) entries.

$$
\begin{array}{c}
N-r \\ N-r+1 \\ N-r+2 \\ \vdots \\ N-r+k \\ N-r+k+1 \\ \vdots \\ N-1 \\ N
\end{array}
\begin{pmatrix}
D_r \\ \tilde{D}_{r-1} \\ \tilde{D}_{r-2} \\ \vdots \\ \tilde{D}_{r-k} \\ D_{r-k-1} \\ \vdots \\ D_1 \\ D_0
\end{pmatrix}
\begin{array}{c} N \\ \\ \\ \\ \\ \\ \\ \\ \end{array}
\qquad (3.14)
$$

$$
\hat{I}_a(b) = \begin{pmatrix} 0 & 0 \\ 0 & I_{b-a} \end{pmatrix}, \quad B_k = \hat{I}_1(N+1) \otimes \hat{I}_k(r) \otimes D_k, \quad 1 \leq k \leq r,
$$
$$
\tilde{B}_k = \hat{I}_1(N+1) \otimes \hat{I}_k(r) \otimes \tilde{D}_k, \quad 1 \leq k \leq r,
$$
$$
\tilde{B}_r = B_r,
$$

(3.15)

and $\tilde{D}_k$, $1 \le k \le r$, is as given in (3.7). Before the generator is displayed, the matrices (in terms of the auxiliary ones) appearing in the generator are defined first:

$$A_i = A_0^{(2)} - i\mu_2 I - \Delta(0, \mu_1 I, 2\mu_1 I, \ldots, N\mu_1 I), \quad 0 \le i \le N - r,$$

$$A_i = \tilde{A}_{i-N+r}^{(2)} - i\mu_2 I - \Delta(0, \mu_1 I, 2\mu_1 I, \ldots, N\mu_1 I), \quad N - r + 1 \le i \le N - 1, \quad (3.16)$$

$$A_N = \tilde{A}_r^{(2)} - N\mu_2 I - \Delta(0, \mu_1 I, 2\mu_1 I, \ldots, N\mu_1 I - \tilde{D}_1).$$

The generator, $Q^{(2)}$ of dimension $m(N+1)^2$, governing the Markov process for the two-SP case is given by

$$Q^{(2)} = \begin{pmatrix} A_0 & B_1 & B_2 & \cdots & B_r & & & & & & \\ \mu_2 I & A_1 & B_1 & B_2 & \cdots & B_r & & & & & \\ & 2\mu_2 I & A_2 & B_1 & B_2 & \cdots & B_r & & & & \\ & & \ddots & \ddots & \ddots & \ddots & \ddots & & & & \\ & & & & \ddots & (N-r)\mu_2 I & & A_{N-r} & \cdots & B_{r-1} & B_r \\ & & & & & & (N-r+1)\mu_1 I & \cdots & \breve{B}_{r-1} \\ & & & & & & \ddots & \ddots & & \\ & & & & & & & & A_{N-1} & \breve{B}_1 \\ & & & & & & & & N\mu_2 I & A_N \end{pmatrix}$$

(3.17)

3.4.1 STEADY-STATE EQUATIONS FOR THE TWO-SERVICE PROVIDER CASE

Let $y = (y_{0,0}, y_{0,1}, \ldots, y_{N,N})$ denote the steady-state probability vector of $Q^{(2)}$. That is, the vector, y, of dimension $m(N+1)^2$ satisfies

$$yQ^{(2)} = 0, \quad ye = 1. \tag{3.18}$$

In terms of vectors and matrices of dimension m suitable for numerical implementation, equation (3.18) can be rewritten as

$$y_{0,0} D_0 + \mu_1 y_{0,1} + \mu_2 y_{1,0} = 0,$$

$$y_{0,j}(D_0 - j\mu_1 I) + \sum_{k=1}^{\min\{j,r\}} y_{0,j-k} D_k + (j+1)\mu_1 y_{0,j+1} + \mu_2 y_{1,j} = 0, \quad 1 \le j \le N-1,$$

$$y_{0,N}(D_0 - N\mu_1 I) + \sum_{k=1}^{r} y_{0,N-k} D_k + \mu_2 y_{1,N} = 0, \tag{3.19}$$

and for $1 \le i \le N - r$,

$$y_{i,0}(D_0 - i\mu_2 I) + \mu_1 y_{i,1} + \mu_2 y_{i+1,0} = 0,$$

$$y_{i,j}(D_0 - j\mu_1 I - i\mu_2 I) + \sum_{k=1}^{\min\{j,r\}} y_{i,j-k} D_k + (j+1)\mu_1 y_{i,j+1} + (i+1)\mu_2 y_{i+1,j}$$

$$= 0, \quad 1 \le j \le N - r,$$

$$y_{i,j}(D_0 - j\mu_1 I - i\mu_2 I) + \sum_{k=1}^{r} y_{i,j-k} D_k + (j+1)\mu_1 y_{i,j+1} + (i+1)\mu_2 y_{i+1,j}$$

$$+ \sum_{k=N-j+1}^{\min\{i,r\}} y_{i-k,j} D_k = 0, \quad N - r + 1 \le j \le N - 1,$$

$$y_{i,N}(D_0 - N\mu_1 I - i\mu_2 I) + \sum_{k=1}^{r} y_{i,N-k} D_k + (i+1)\mu_2 y_{i+1,N} + \sum_{k=1}^{\min\{i,r\}} y_{i-k,N} D_k = 0,$$

$$(3.20)$$

and for $N - r + 1 \le i \le N - 1$,

$$y_{i,0}(D_0 - i\mu_2 I) + \mu_1 y_{i,1} + \mu_2 y_{i+1,0} = 0,$$

$$y_{i,j}(D_0 - j\mu_1 I - i\mu_2 I) + \sum_{k=1}^{\min\{j,r\}} y_{i,j-k} D_k + (j+1)\mu_1 y_{i,j+1} + (i+1)\mu_2 y_{i+1,j}$$

$$= 0, \quad 1 \le j \le N - r,$$

$$y_{i,j}(D_0 - j\mu_1 I - i\mu_2 I) + \sum_{k=1}^{r} y_{i,j-k} D_k + (j+1)\mu_1 y_{i,j+1} + (i+1)\mu_2 y_{i+1,j}$$

$$+ \sum_{k=N-j+1}^{\min\{i,r\}} y_{i-k,j} D_k = 0, \quad N - r + 1 \le j \le N - 1,$$

$$y_{i,N}(D_0 - N\mu_1 I - i\mu_2 I) + \sum_{k=1}^{r} y_{i-k,N} D_k + (i+1)\mu_2 y_{i+1,N} + \sum_{k=1}^{N-i-1} y_{i,N-k} D_k$$

$$+ \sum_{k=N-i}^{r} y_{i,N-k} \tilde{D}_k = 0, \quad (3.21)$$

and

$$y_{N,0}(D_0 - N\mu_2 I) + \mu_1 y_{N,1} = 0,$$

$$y_{N,j}(D_0 - j\mu_1 I - N\mu_2 I) + \sum_{k=1}^{\min\{j,r\}} y_{N,j-k} D_k + (j+1)\mu_1 y_{N,j+1} = 0, \ 1 \le j \le N - r,$$

$$y_{N,j}(D_0 - j\mu_1 I - N\mu_2 I) + \sum_{k=1}^{r} y_{N,j-k} D_k + (j+1)\mu_1 y_{N,j+1} + \sum_{k=N-j+1}^{\min\{i,r\}} y_{i-k,j} \tilde{D}_k$$

$$= 0, \ N - r + 1 \le j \le N - 1,$$

$$y_{N,N}(D_0 - N\mu_1 I - N\mu_2 I) + \sum_{k=1}^{r} y_{N-k,N} \tilde{D}_k + \sum_{k=1}^{r} y_{N,N-k} \tilde{D}_k = 0, \quad (3.22)$$

subject to the normalizing condition

$$\sum_{i=0}^{N} \sum_{j=0}^{N} y_{i,j} e = 1. \tag{3.23}$$

Like in the one-*SP* case, it is easy to verify that

$$\sum_{i=0}^{N} \sum_{j=0}^{N} y_{i,j} = \pi, \tag{3.24}$$

where π is as given in (3.1). This will also serve as another accuracy check in numerical implementation of the computation of y.

3.4.2   SYSTEM MEASURES FOR TWO-SERVICE PROVIDER CASE

In this section, a few key performance measures to qualitatively describe and compare various models studied in this chapter are listed along with their formulas.

(1) The idle probability, P_{idle_1}, of SP_1, is given by

$$P_{idle} = \sum_{i=0}^{N} y_{i,0} e.$$

(2) The idle probability, P_{idle_2}, of SP_2, is given by

$$P_{idle} = \sum_{j=0}^{N} y_{0,j} e.$$

(3) The mean number in the system (which is also the mean number of busy resources), μ_{NS}, is calculated as

$$\mu_{NS} = \mu_{NS_1} + \mu_{NS_2} = \sum_{j=1}^{N} j \sum_{i=0}^{N} y_{i,j} e + \sum_{i=1}^{N} i \sum_{j=0}^{N} y_{i,j} e.$$

(4) The correlation coefficient of the number busy in SP_1 and SP_2 is obtained as

$$\rho_{NS} = \frac{\sum_{i=0}^{N} \sum_{j=0}^{N} ij y_{i,j} e - \mu_{NS_1} \mu_{NS_2}}{\sigma_{NS_1} \sigma_{NS_2}},$$

where σ_{NS_i} is the standard deviation of the number busy in SP_i, $i = 1, 2,$, which is computed similar to the one-*SP* case, and the details are omitted here.

(5) The probability, P_{loss}, that a loss occurs at an arrival epoch, is obtained as

$$P_{loss} = \frac{1}{\lambda_g} \sum_{k=1}^{r} \sum_{j=k}^{r} \left[\sum_{i=N-r+j}^{N} \left(y_{i,N-r+j} + y_{N-r+j,i} \right) - y_{N-r+j,N-r+j} \right] \tilde{D}_{r-j+k} e.$$

(6) The mean number of lost items, μ_{loss}, at an arrival epoch, is calculated as

$$\mu_{loss} = \frac{1}{\lambda_g} \sum_{k=1}^{r} \sum_{j=k}^{r} k \left[\sum_{i=N-r+j}^{N} \left(y_{i,N-r+j} + y_{N-r+j,i} \right) - y_{N-r+j,N-r+j} \right] \tilde{D}_{r-j+k} e.$$

(7) The throughput, γ, which is the rate at which the items are processed per unit of time, is computed as

$$\gamma = \lambda - \lambda_g \mu_{loss}.$$

3.5 SYSTEMS WITH MORE THAN TWO-SERVICE PROVIDERS

While one can study the systems having three or more *SPs* along the lines of the previous sections, here simulation approach is taken for the following reasons. First, the state space of the models increases exponentially as the number of *SPs* is increased. That is, if there are n *SPs* and assuming all these *SPs* have an identical number of resources, then the state space will be of dimension $m(N+1)^n$. Second, one may wish to look at various variations such as services to be non-exponential, which might further increase the dimensionality of the problem. This, along with other variations, will be elaborated later.

Before the simulation models are used for qualitative interpretation, it is very important to validate the simulated ones against the analytical models. Toward this end, the simulated models are validated against the one- and two-*SP* cases discussed (analytically) in earlier sections. The simulation is conducted using ARENA (see, e.g., Kelton, Sadowski, & Swets (2010)). In Figure 3.1, the simulation model used for a four-*SP* case is displayed. Under a variety of scenarios, the simulated ones are compared against the corresponding analytical models. The error percentages, which are calculated as the ratio of the absolute differences in the simulated and analytical values over the analytical values, varied anywhere between 0.016% and 5.722%. This indicates that the simulated models appear to yield results that are consistent with the analytical models. The simulated results are presented along with the analytical ones in the next section.

3.6 ILLUSTRATIVE EXAMPLES

In this section, selected illustrative examples to bring out the qualitative behavior of the models (analytical models for one- and two-*SP* cases, and simulated models for three- and four-*SP* cases) are presented.

In all our illustrative numerical examples below, 16 different *BMAPs*, of which 8 have negatively correlated arrivals and the other 8 have positively correlated arrivals, are used. The full details of these (in the context of single arrivals) are given in Chakravarthy (2021) and Chakravarthy (2020b). However, for the sake of completeness and for our consideration here in the context of batch arrivals, some key items are briefly outlined.

Let p_k, $1 \leq k \leq r$, denote the probability that an arriving batch is of size k. The *BMAP* representation, $\{D_k^{(i)}\}$, of dimension $i+2, 1 \leq i \leq 8$, of the eight negatively

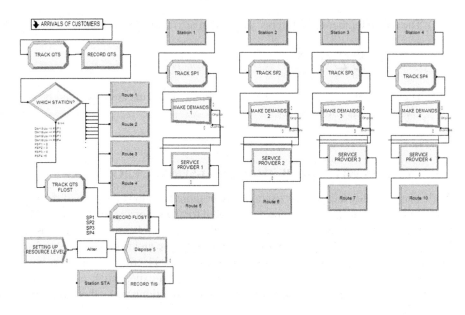

Figure 3.1 ARENA simulation model for a four-service provider case.

correlated arrivals denoted as NC_i and the eight positively correlated arrivals denoted as PC_i are displayed below.

1. *BMAP* **with negative correlation** $(NC_i,\ 1 \leq i \leq 8)$**:**

$$D_0^{(i)} = (0.5i + 0.75) \begin{pmatrix} -1 & 1 & 0 & \cdots & 0 & 0 \\ 0 & -1 & 1 & \cdots & 0 & 0 \\ \vdots & \vdots & \vdots & \cdots & \vdots & \vdots \\ 0 & 0 & 0 & \cdots & -1 & 0 \\ 0 & 0 & 0 & \cdots & 0 & -2 \end{pmatrix},$$

$$D_k^{(i)} = (0.5i + 0.75)p_k \begin{pmatrix} 0 & 0 & 0 & \cdots & 0 & 0 \\ 0 & 0 & 0 & \cdots & 0 & 0 \\ \vdots & \vdots & \vdots & \cdots & \vdots & \vdots \\ 0.01 & 0 & 0 & \cdots & 0 & 0.99 \\ 1.98 & 0 & 0 & \cdots & 0 & 0.02 \end{pmatrix},\ k \geq 1.$$

$$(3.25)$$

2. *BMAP* **with negative correlation** $(PC_i, \ 1 \leq i \leq 8)$:

$$D_0^{(i)} = (0.5i + 0.75) \begin{pmatrix} -1 & 1 & 0 & \cdots & 0 & 0 \\ 0 & -1 & 1 & \cdots & 0 & 0 \\ \vdots & \vdots & \vdots & \cdots & \vdots & \vdots \\ 0 & 0 & 0 & \cdots & -1 & 0 \\ 0 & 0 & 0 & \cdots & 0 & -2 \end{pmatrix},$$

$$D_k^{(i)}(i) = (0.5i + 0.75)p_k \begin{pmatrix} 0 & 0 & 0 & \cdots & 0 & 0 \\ 0 & 0 & 0 & \cdots & 0 & 0 \\ \vdots & \vdots & \vdots & \cdots & \vdots & \vdots \\ 0.99 & 0 & 0 & \cdots & 0 & 0.01 \\ 0.02 & 0 & 0 & \cdots & 0 & 1.98 \end{pmatrix}, \ 1 \leq k \leq r.$$

$$(3.26)$$

The above *BMAPs* will be normalized so as to have the same λ_g, which is taken to be 10 in this chapter. These *BMAPs* are qualitatively different. To illustrate this, the graphs of the probability density function and the joint probability density function of the two *BMAPs*, namely, NC_1, NC_8, PC_1, and PC_8 are plotted in Figure 3.2. The one lag correlation coefficient, $\hat{\rho}$, is displayed on the plot itself. It is clear from this plot how qualitatively these *BMAPs* differ. The standard deviation (σ), which is the same for both NC_i and PC_i and the one lag correlation coefficient values of the above 16 *BMAPs* as adapted from Chakravarthy (2021) are given in Table 3.1.

In the following examples, a truncated Poisson distribution for $\{p_k\}$ is considered. Note that as mentioned earlier, the maximum batch size is taken to be r. Also, it is worth pointing out that uniform and geometric distributions for the batch size were also considered, and the patterns (to be discussed below) noticed in the performance measures were similar to that of the truncated Poisson. Thus, the focus here is only on one scenario, namely, the truncated Poisson with regard to the batch size distribution.

Truncated Poisson Batch Size The arriving batch is of size k with probability given by $p_k = e^{-\theta} \left(\dfrac{\theta^{k-1}}{(k-1)!} \right)$, $1 \leq k \leq r-1$. The value of $p_r = 1 - \sum_{k=1}^{r-1} p_k$, which makes the probability mass function on finite support to be a truncated Poisson. Further, the parameter θ is chosen such that the mean batch size is $0.5(r+1)$. This restriction is mainly to compare with other batch size distributions such as uniform and truncated geometric. As pointed out earlier, similar patterns were seen in these cases too, and due to space restriction, only truncated Poisson case is discussed.

In this chapter, the following values for r, the maximum batch size: $r = 3, 5, 7, 9$ are considered. The numerical examples are discussed as follows. First, the measures that are unique to one- and two-*SP* systems are discussed. The scenarios which require comparisons with various other *SP* cases are presented later.

EXAMPLE 3.1 (One-service provider case based on the analytical model): In this example, fixing $\mu_1 = 0.5$, the optimum, namely, the minimum number of resources, say, N^*, of N so that the throughput is at least 99% of the input rate of $\lambda = 0.5(r+1)\lambda_g$ is obtained for a single-*SP* (SP_1) case. This way, the number of

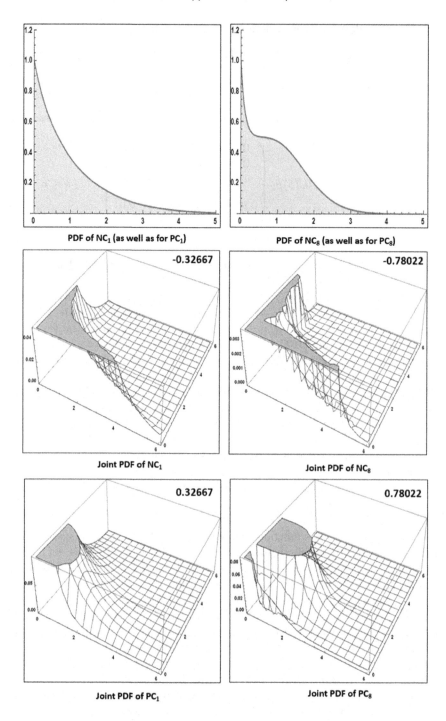

Figure 3.2 PDF and joint PDF of selected *BMAPs* (with 1-lag correlation displayed).

Table 3.1
BMAPs along with Their $\hat{\rho}$ and σ

BMAP	$\hat{\rho}$	BMAP	$\hat{\rho}$	σ
NC_1	−0.3267	PC_1	0.3267	0.10392
NC_2	−0.4804	PC_2	0.4804	0.10202
NC_3	−0.5786	PC_3	0.5786	0.10123
NC_4	−0.6454	PC_4	0.6454	0.10082
NC_5	−0.6935	PC_5	0.6935	0.10059
NC_6	−0.7296	PC_6	0.7296	0.10044
NC_7	−0.7577	PC_7	0.7577	0.10035
NC_8	−0.7802	PC_8	0.7802	0.10028

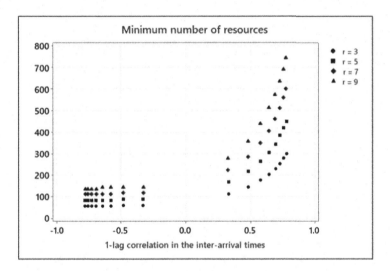

Figure 3.3 Minimum resources vs 1-lag correlation.

resources identified will be adequate to ensure that the loss of customers is restricted to be no more than 1% of the arrival rate per unit of time. In Figure 3.3, the values of N^* under various scenarios are plotted.

It is very clear from Figure 3.3 that

- when going from NC_8 to NC_1 (which corresponds to the correlation going from −0.7802 to −0.3267), one notices that N^* shows a non-decreasing trend, even if the rate of change (in N^*) is not that significant. This is true for $r = 3, 5, 7, 9$. Further, as is to be expected, one needs a large value of N^* when r increases. That is, N^* increases as r is increased.
- in the case of positively correlated arrival processes, while an increasing trend in N^* is seen as the correlation increases from 0.3267 to 0.7802, the

rate of increase is significantly high. For example, when $r = 3$, going from a correlated value of 0.3267 to 0.7802, N^* goes from 114 to 300. Again, as is to be expected N^* increases as r is increased. Further, the rate of increase is also significantly higher.

Now, the mean, the mode, and selected percentiles under various scenarios are displayed in Figure 3.4. Some interesting observations from this figure are registered as follows.

- The mean, the median, and the mode of the number of busy resources appear to be close to each other in the case of negatively correlated arrivals, indicating a symmetric nature of the probability mass function of the number of busy resources. Further, the mean, the median, and the mode appear to be insensitive to the type of negatively correlated arrivals, whereas some sensitivity is seen with regard to a few other measures, namely, 10th and 95th percentiles. This indicates that depending on the application one should not ignore negative correlation in the arrival process by looking only at the standard measures like the mean (and at times the variation). Overlooking the correlation has been very prevalent in the literature. All the measures are sensitive to the value of r, as is to be expected.
- With regard to positively correlated arrivals, a highly skewed (to the right) distribution for the number of busy resources is noticed. Further, only the mean appears to be insensitive to the type of positively correlated arrivals. However, the mean is sensitive to r in that it increases as r increases. With respect to the other measures, notice that (in all batch sizes considered) as the correlation increases (i.e., going from PC_1 to PC_8) (i) the mode, the median, 10th, 30th, 70th percentiles all decrease; and (ii) the 95th percentile increases.

EXAMPLE 3.2 (Two-service provider case based on the analytical model): In this example, taking $\mu_1 = \mu_2 = 0.5$ a two-SP case is considered. The number of resources identified in Example 3.1 (in the case of the one-SP system) is split equally between the two SPs. For example when looking at the arrival process to be NC_1 with $r = 3$, the value of N found to be 60 in Example 1 will be split as 30 for SP_1 and 30 for SP_2. The rationale for using the total resources to be 60 is to properly compare the key measures between one and two-SP cases. While one can split this 60 in a variety of ways between SP_1 and SP_2, an equal split is considered here. More on this to be discussed in Section 3.7.

In Figures 3.5 and 3.6, respectively, the joint probability mass function of the number of resources in SP_1 and SP_2 busy for selected scenarios are displayed. These figures clearly indicate the significant differences in the behavior of the negatively correlated and the positively correlated arrivals.

In Table 3.2, the value of N, the values of (k_1, k_2) where the joint probability mass function of the number of busy resources in (SP_1, SP_2) peaks, and the corresponding joint probability (ξ) are displayed. Thus, the joint probability function has a mode at (k_1, k_2) and the value at that point is denoted by ξ.

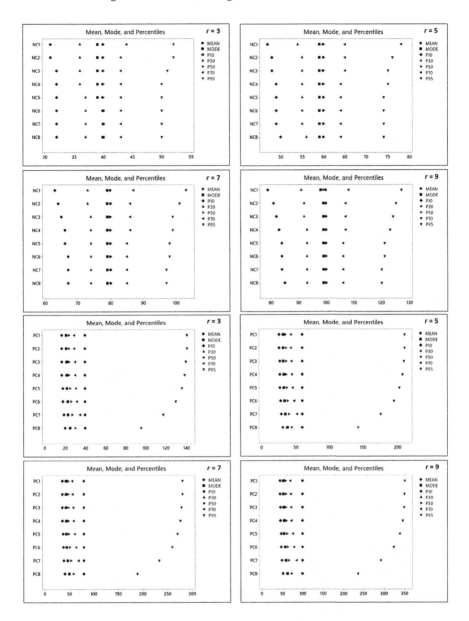

Figure 3.4 Selected measures for one-service provider case under various scenarios.

It is clear from the values displayed in Table 3.2 that as r increases, ξ tends to decrease for both negative and positively correlated cases. Further, as the magnitude of correlation increases ξ appears to increase, and once again indicates the role of correlation in the arrival process. A very interesting aspect is that for the positively

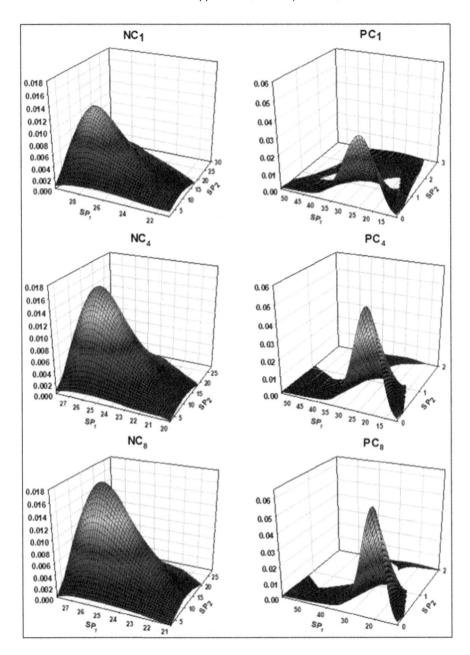

Figure 3.5   Joint probability mass function under various scenarios when $r = 3$.

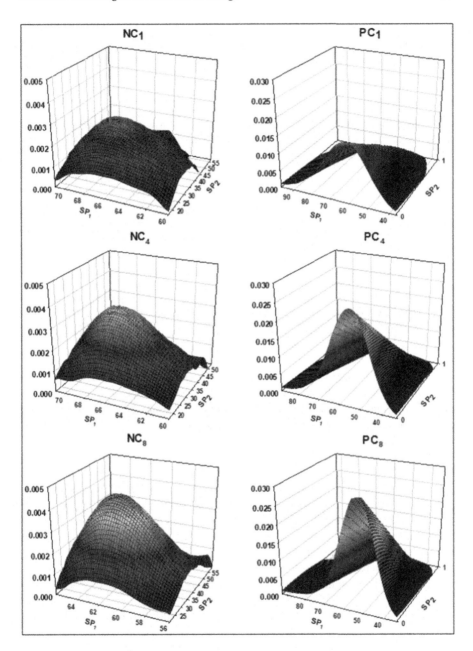

Figure 3.6 Joint probability mass function under various scenarios when $r = 9$.

Table 3.2

$(N, k_1, k_2, \xi))$ **when** $\lambda = 10, \mu = 0.5$ **under Various Scenarios**

TAP	$r = 3$	$r = 5$	$r = 7$	$r = 9$
NC_1	(30, 28, 13, 0.01147)	(45, 42, 19, 0.00583)	(59, 55, 27, 0.00366)	(72, 67, 35, 0.00258)
NC_2	(30, 28, 13, 0.01265)	(45, 42, 19, 0.00653)	(59, 55, 26, 0.00413)	(72, 67, 35, 0.00292)
NC_3	(28, 27, 15, 0.01436)	(42, 39, 22, 0.00746)	(56, 52, 29, 0.00470)	(72, 67, 34, 0.00318)
NC_4	(28, 27, 14, 0.01493)	(42, 39, 21, 0.00785)	(56, 52, 29, 0.00497)	(72, 67, 34, 0.00339)
NC_5	(28, 27, 14, 0.01539)	(42, 39, 21, 0.00815)	(56, 52, 29, 0.00518)	(67, 62, 38, 0.00377)
NC_6	(28, 27, 14, 0.01574)	(42, 39, 21, 0.00839)	(56, 52, 29, 0.00536)	(67, 62, 38, 0.00391)
NC_7	(28, 27, 14, 0.01604)	(42, 39, 21, 0.00860)	(56, 52, 29, 0.00550)	(67, 62, 38, 0.00402)
NC_8	(28, 27, 14, 0.01628)	(42, 39, 21, 0.00877)	(56, 52, 28, 0.00562)	(67, 62, 38, 0.00412)
PC_1	(57, 25, 0, 0.03995)	(85, 37, 0, 0.02748)	(112, 49, 0, 0.02098)	(139, 61, 0, 0.01700)
PC_2	(73, 23, 0, 0.04831)	(110, 35, 0, 0.03377)	(143, 46, 0, 0.02616)	(180, 58, 0, 0.02147)
PC_3	(88, 22, 0, 0.05336)	(132, 33, 0, 0.03774)	(175, 45, 0, 0.02957)	(220, 56, 0, 0.02453)
PC_4	(102, 22, 0, 0.05702)	(152, 33, 0, 0.04058)	(203, 44, 0, 0.03202)	(256, 55, 0, 0.02669)
PC_5	(115, 22, 0, 0.05934)	(172, 32, 0, 0.04263)	(231, 43, 0, 0.03389)	(288, 54, 0, 0.02834)
PC_6	(127, 21, 0, 0.06134)	(192, 32, 0, 0.04438)	(255, 43, 0, 0.03529)	(319, 53, 0, 0.02959)
PC_7	(139, 21, 0, 0.06308)	(210, 32, 0, 0.04568)	(280, 42, 0, 0.03642)	(346, 53, 0, 0.03067)
PC_8	(150, 21, 0, 0.06443)	(225, 32, 0, 0.04664)	(301, 42, 0, 0.03741)	(373, 53, 0, 0.03152)

correlated arrivals, it appears that SP_2 is idle more often than being busy in any particular state (even though the total busy probability is larger than the idle probability of SP_2). It is intuitively clear that the peak occurs at points where more resources are busy in SP_1 as compared to SP_2. This is due to the way SP_1 is used more often than SP_2.

In Figure 3.7, the correlation, say, ζ, between the numbers busy in SP_1 and SP_2 is plotted. First note that ζ is positive under all scenarios indicating that the number of resources busy in SP_2 increase with an increase in SP_1 but the rate of increase depends on the type of (correlated) arrivals. With respect to negatively correlated arrivals, one can see that ζ increases when going from NC_1 to NC_8. That is, when the magnitude of the (negative) correlation increases, ζ decreases. However, when looking at positively correlated arrivals, a slightly different pattern is noticed. While ζ appears to increase initially (when going from PC_1 to PC_2), it starts to decrease. In any case, when the magnitude of correlation is large, the value of ζ appears to decrease.

EXAMPLE 3.3 (Multiservice provider case based on simulated models): In this example, the simulated models of three- and four-SP cases along with the analytical models of the one and two-SP cases are used to bring out the qualitative nature of the models studied in this chapter. Towards this end, the measures (i) utilization and (ii) coefficient of variation of the number of busy resources are used. So, for proper comparisons, the service rates are fixed as $\mu_1 = \mu_2 = \mu_3 = \mu_4 = 0.5$, and the number of resources (identified in Example 3.1) is split equally among the various SP systems. Thus, if $N = 60$ for the one-SP system, then the number of resources for

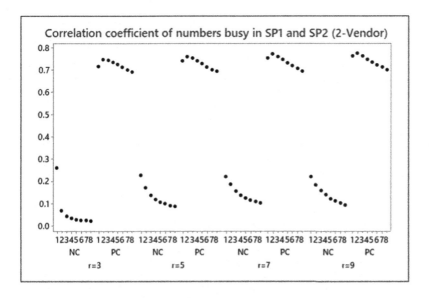

Figure 3.7 Correlation of numbers busy in SP_1 and SP_2 in the two-service provider case.

the three-*SP* system is 20 for each of the three systems, and for the four-*SP* systems it will be 15 for each of the four. If N is not a proper multiple of 3 or 4, it will be rounded down. For example, if $N = 40$ and a three-*SP* is considered, then each of these three *SPs* will have 13 resources each. It should be pointed out only in some scenarios the throughput was about 98% as opposed to 99% of the actual ones. This could be due to the sampling errors in the simulation.

The two measures, the utilization and the coefficient of variation of the number of resources busy of various *SPs* are plotted in Figures 3.8–3.11 under different scenarios.

An examination of these figures reveals the following:

- While a non-decreasing trend in the utilization of the resources in SP_1 system when the arrival process is varied from NC_1 to NC_8 is noticed, a decreasing trend in this measure for the SP_1 system is seen when the arrival process is varied from PC_1 to PC_8. This is the case for all one-*SP*, two-*SP*, three-*SP* and four-*SP* systems, and for all values of r considered.
- The utilization for the negatively correlated arrivals is significantly higher than the corresponding scenario for the positively correlated arrivals.
- The coefficient of variation for the positively correlated arrivals is significantly higher than the corresponding scenario for the negatively correlated arrivals. This shows that positively correlated arrivals exhibit more erratic behavior than the negatively correlated arrivals. It should be mentioned that NC_i and PC_i arrivals have the same standard deviation.

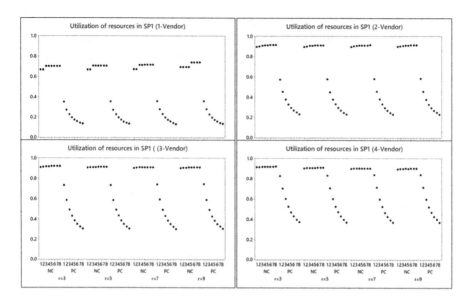

Figure 3.8   Comparison of utilization in SP_1 among four-service providers.

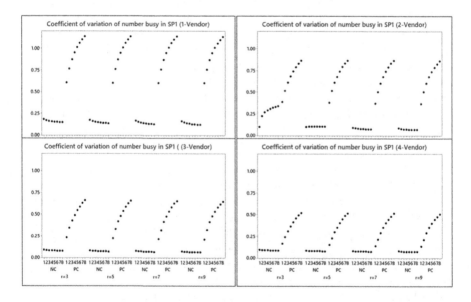

Figure 3.9   Comparison of CV in SP_1 among four-service providers.

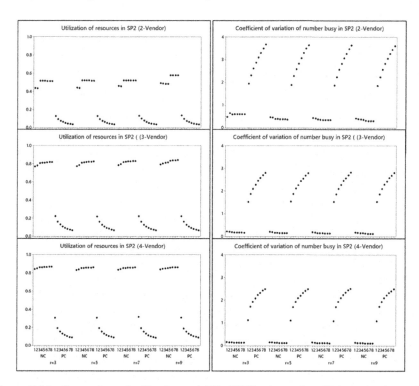

Figure 3.10 Comparison of utilization and CV of Utilization and coefficient of variation for SP2 under various scenarios.

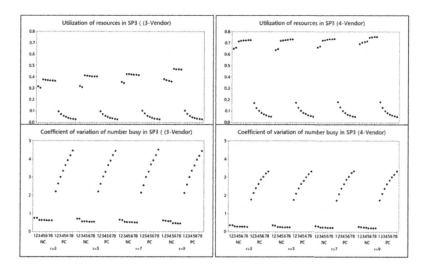

Figure 3.11 Utilization and coefficient of variation for SP3 under various scenarios.

3.7 CONCLUDING REMARKS

In this chapter, stochastic models useful in service-related items in the context of *IoT* are studied. The demand (or arrival) processes are modeled using Neuts' versatile Markovian point process. After identifying the minimum number of resources (or servers) required to maintain the throughput level to be close to the input rate (due to having a finite but sufficiently large resources in the system), one- and two-*SP* systems are analyzed analytically using matrix-analytic methods. A simulation approach is taken for three-*SP* and four-*SP* systems. The quantification of the number of needed resources coupled with pointing out the significant roles of the correlated (negative as well as positive) arrivals, will enable practicing managers to allocate resources accordingly. While this chapter focused on the assumption of batch Markovian arrivals and exponential services with a finite number of resources among one or more *SP* systems, there remains more studies along the lines outlined here. The main purpose of this chapter is to introduce such studies in the context of *IoT*, which is going to continue to permeate our lives in the decades to come.

The model studied in this chapter can be extended in many forms as follows. (i) While the models focused only on partial admissions due to lack of available resources, it is easy to modify to consider full admissions situation. Instead of summing the D_k matrices in the last (block) column (see, e.g., 3.8) as needed for the partial admissions, one would sum these along their respective diagonals for one-*SP* system. A similar modification is needed for other systems. It would be of interest to compare corresponding full and partial admission control policies. (ii) The assumption of exponential services can be relaxed to include phase type services. However, in this case due to the number of resources even having a common phase type services will increase the dimension of the problem exponentially. One way to overcome this difficulty is to adapt some form of a threshold policy, such as the one introduced recently in the context of a retrial queue Chakravarthy (2020a). (iii) The assumption of the *SP* accepting (subject to the availability of the resources) all items submitted to them can be modified to allow the possibility of not accepting for a variety of reasons. This calls for the introduction of a Bernoulli parameter. (iv) The assignment of a *SP* in the case of a multiservice provider system can be looked upon from different angles than the one studied here. For example, one can assign a *SP* based on the least number of resources busy. (v) It is assumed that a batch of items cannot be split and sent to more than one-*SP*. However, that can be relaxed and identify a scheme that will further reduce the average number of losses per unit of time. (vi) While the models did not allow any demand to be queued in any *SP* area, one can modify this assumption. (vii) When looking at more than one *SP* case, the resources are split equally among them. It would be of interest to look at different ways to split. This relaxation calls for using some clever schemes due to the number of combinations of such configurations one has to consider. (viii) One can introduce a monitoring system in *CRU* such that dynamic decisions can be made. For example, decisions to move the items from one *SP* to another or entirely cancel the item, can be made based on having timers possibly one per *SP* where there are unfulfilled items. Each timer will be given a different set of initial conditions. The idea is that once the timer goes off, that particular *SPs* items will be evaluated for further processing. The actions will include, canceling the item placed with a particular *SP* and reordering the same item(s) from another *SP*, deciding to wait for some additional time by re-initializing the timer of that *SP*, or altogether canceling the item without placing a reorder due mainly to lack of the availability of items, or lack of timely arrival of information. Finally, simulation will play a key role, especially when the model assumptions significantly increase the dimension of the problem, and in such cases, a thorough study requires very careful planning and collecting of data for practical implementation.

REFERENCES

Alfa, A. S. (2010). *Queueing Theory for Telecommunications*. Springer Science+Business Media, LLC.

Artalejo, J.R., Gomez-Correl, A., and He, Q. M. (2010) Markovian arrivals in stochastic modelling: a survey and some new results. *SORT*, 34(2), 101–144.

Ashton, K. (2009). That 'Internet of Things' Thing, In the real world, things matter more than ideas. Online: http://www.rfidjournal.com/articles/view?4986.

Balas, V.E., Solanki, V.K., Kumar, R., and Ahad, Md. (Eds.). (2020). *A Handbook of Internet of Things in Biomedical and Cyber Physical System*. Springer Nature, Switzerland AG.

Balas, V.E., Solanki, V.K., Kumar, R., and Srivastava, R. (Eds.). (2020). *Recent Trends and Advances in Artificial Intelligence and Internet of Things*. Springer Nature, Switzerland AG.

Bhat, U. N. (2015). *An Introduction to Queueing Theory*. Second Edition, Springer Science+Business Media, New York.

Chakravarthy, S. R. (2001). The batch Markovian arrival process: A review and future work. *Advances in Probability Theory and Stochastic Processes*. A. Krishnamoorthy et al. (Eds.), Notable Publications Inc., NJ, 21–39.

Chakravarthy, S. R. (2010). Markovian arrival processes. Wiley Encyclopedia of Operations Research and Management Science. Published Online: 15 JUN 2010.

Chakravarthy, S. R. (2015). Matrix-Analytic Queueing Models, Chapter 8. *An Introduction to Queueing Theory*. U. Narayan Bhat (Ed.), Second Edition, Birkhauser, Springer Science + Business Media, New York.

Chakravarthy, S. R. (2020). A retrial queueing model with thresholds and phase type retrial times. *Journal of Applied Mathematics and Informatics*, 38, 3–4, 351–373.

Chakravarthy, S. R. (2020). Busy period analysis of multi-server retrial queueing systems. *Applied Probability and Stochastic Processes*, V. C. Joshua et al. (Eds.), Infosys Science Foundation Series, pp. 61–76.

Chakravarthy, S.R. (2021). Modèles de files d'attente dans les services – approche analytique et de simulation. *Théorie des files d'attente 2*, Prof. Vladimir Anisimov and Prof. Nikolaos Limnios. (Eds.), Series of books "Mathematics and Statistics Sciences, ISTE Editions Ltd 2021.

Chindanonda, P., Podolskiy, V., and Gerndt, M. (2019). Metrics for Self-Adaptive Queuing in Middleware for Internet of Things. IEEE 4th International Workshops on Foundations and Applications of Self* Systems (FAS*W), Umea, Sweden, pp. 130–133.

Dudin, A. N., Klimenok, V. I., and Vishnevsky, V. M. (2020). *The Theory of Queueing Systems with Correlated Flows*. Springer Nature, Switzerland AG.

Evans. D. (2011). The Internet of Things: How the next evolution of the Internet is changing everything, white paper, Cisco.

Firouzi, F., Chakrabarty, K., and Nassif, S. (Eds.). (2020). *Intelligent Internet of Things from Device to Fog and Cloud*. Springer Nature, Switzerland AG.

Graham, A. (1981). *Kronecker Products and Matrix Calculus with Applications*. Ellis Horwood, Chichester, UK.

Harchol-Balter, M. (2013). *Performance Modeling and Design of Computer Systems: Queueing Theory in Action*. Cambridge University Press.

Hassija, V. Chamola, V. Saxena, D. Jain, P. Goyal, and B. Sikdar. (2019). A Survey on IoT Security: Application Areas, Security Threats, and Solution Architectures. *IEEE Access*, 7, 82721–82743.

He, Qi-Ming. (2014). *Fundamentals of Matrix-Analytic Methods*. Springer, New York.

Kanagachidambaresan, G.R., Anand, R., Balasubramanian, E., and Mahima, V. (Eds.). (2020). *Internet of Things for Industry 4.0 Design, Challenges and Solutions*. Springer Nature, Switzerland AG.

Kelton, W. D., Sadowski, R. P., Swets, N. B. (2010). *Simulation with ARENA*, Fifth edition, McGraw-Hill, New York.

Kua, J., Nguyen, S. H., Armitage, G., and Branch, P. (2017). Using active queue management to assist IoT application flows in home broadband networks. *IEEE Internet of Things Journal*, 4 (5), 1399–1407.

Li, J., Zaho, Y.Q., Yu, R., and Huang, X. (2016). Queuing analysis of two-hop relay technology in LTE/LTE-A networks with unsaturated and asymmetric traffic. *IEEE Internet of Things Journal*, 3 (3), 378–385.

Lucantoni, D. M., Meier-Hellstern, K. S., and Neuts, M. F. (1990). A single-server queue with server vacations and a class of nonrenewal arrival processes. *Advances in Applied Probability*, 22, 676–705.

Lucantoni, D. M. (1991). New results on the single server queue with a batch Markovian arrival process. *Stochastic Models*, 7, 1–46.

Mahmood, Z. (Ed.). (2019). *The Internet of Things in the Industrial Sector Security and Device Connectivity, Smart Environments, and Industry 4.0*. Springer Nature, Switzerland AG.

Mahmood, Z. (Ed.). (2020). *Connected Vehicles in the Internet of Things Concepts, Technologies and Frameworks for the IoV*. Springer Nature, Switzerland AG.

Marcus, M. and Minc, H. (1964). *A Survey of Matrix Theory and Matrix Inequalities*. Allyn and Bacon, Boston, MA.

Milenkovic, M. (2020). *Internet of Things: Concepts and System Design*. Springer Nature, Switzerland AG, 2020.

Mohamed, K.S. (2019). *The Era of Internet of Things towards a Smart World*. Springer Nature, Switzerland AG.

Neuts, M. F. (1975). Probability distributions of phase type. In *Liber Amicorum Prof. Emeritus H. Florin*, Department of Mathematics, University of Louvain, 173–206.

Neuts, M. F. (1979). A versatile Markovian point process. *Journal of Applied Probability*, 16, 764–779.

Neuts, M. F. (1981). *Matrix-Geometric Solutions in Stochastic Models: An Algorithmic Approach*. The Johns Hopkins University Press, Baltimore, MD. [1994 version is Dover Edition].

Neuts, M. F. (1989). *Structured Stochastic Matrices of $M/G/1$ Type and Their Applications*. Marcel Dekker, Inc., New York.

Neuts, M. F. (1992). Models based on the Markovian arrival process. *IEICE Transactions on Communications*, E75B, 1255–1265.

Neuts, M. F. (1995). *Algorithmic Probability: A Collection of Problems*. Chapman and Hall, New York.

Qiao, R., Qiu, T., Han, M., Ma, J., and Huang, R.(2017). An Emergent Backpressure Queueing Model for Internet of Things. 2017 IEEE International Conference on Internet of Things (iThings) and IEEE Green Computing and Communications (GreenCom) and IEEE Cyber, Physical and Social Computing (CPSCom) and IEEE Smart Data (SmartData), Exeter, 651–654.

Raj, P., Chatterjee, J. M., Kumar, A., and Balamurugan, B. (Eds.). (2020). *Internet of Things Use Cases for the Healthcare Industry*. Springer Nature, Switzerland AG.

Rathod, D. and Chowdhary, G. (2019). Scalability of M/M/c queue based cloud-fog distributed Internet of Things middleware. *International Journal of Advance in Networking and Applications*, 11 (1), 4162–4170.

Rayes, A. and Salam, S. (2019). *Internet of Things from Hype to Reality - The Road to Digitization*, Second Edition. Springer Nature, Switzerland AG.

Salameh, O., Awad, M., and AbuAlrub, F. (2019). A Markovian model for Internet of Things. it International Journal of Computer Networks & Communications, 11(2), 113-124.

Sharma, R., Kumar, N., Gowda, N. B., and Srinivas, T. (2015). Waiting Time Analysis for Delay Sensitive Traffic in Internet of Things. 2015 IEEE Region 10 Symposium, Ahmedabad, pp. 58–61.

Steeb, W. H. and Hardy, Y. (2011). *Matrix Calculus and Kronecker Product*, World Scientific Publishing, Singapore.

Strielkina, T., Uzun, D., and Kharchenko, V. (2017). Modelling of healthcare IoT using the queueing theory. 9th IEEE International Conference on Intelligent Data Acquisition and Advanced Computing Systems: Technology and Applications (IDAACS), Bucharest, 2017, pp. 849–852.

Viriyavisuthisakul, S., Sanguansat, P., Toriumi, S., Hayashi, M., and Yamasaki, T. (2017). Automatic queue monitoring in store using a low-cost IoT sensing platform., 2017 IEEE International Conference on Consumer Electronics - Taiwan (ICCE-TW), Taipei, pp. 53–54.

Volochiy, B., Yakovyna, V., and Mulyak, O.v(2017). Queueing networks for availability and safety assessment of the IoT data service. 2017 12th International Scientific and Technical Conference on Computer Sciences and Information Technologies (CSIT), Lviv, 2017, pp. 393–396.

Yousefpour, A., Ishigaki,G., Gour, R., and Jue, J.P. (2018). On reducing IoT service delay via fog offloading. *IEEE Internet of Things Journal*, 5(2), 998–1010.

4 FTLB: An Algorithm for Fault Tolerant Load Balancing in Fog Computing

Ashish Virendra Chandak and Niranjan Kumar Ray
Kalinga Institute of Industrial Technology, Deemed to be University
Bhubaneswar, India

CONTENTS

4.1   Introduction 66
4.2   Background Details 67
     4.2.1   Existing Fog Computing Architecture 67
     4.2.2   Existing Load Balancing Strategies 69
     4.2.3   Existing Fault Tolerance Strategies 69
4.3   Problem Statement, Notations, and Definitions 70
     4.3.1   Problem Statement 70
     4.3.2   Notations and Definitions 71
4.4   Proposed Model for Load Balancing and Fault Tolerance 71
     4.4.1   Fog Computing Environment 71
     4.4.2   Components of Proposed Model 73
     4.4.3   Algorithm Description 74
          4.4.3.1   Pseudocode of FTLB Algorithm 75
4.5   A Brief Description of Load Balancing Algorithms 75
     4.5.1   Round Robin 75
     4.5.2   Weighted Round Robin 76
     4.5.3   Random Algorithm 76
4.6   Experiments 76
     4.6.1   Simulation Environment 76
     4.6.2   Performance Analysis 78
4.7   Conclusion 83
References 83

4.1 INTRODUCTION

Today, there is the worldwide adoption of smart services, and smart services are being implemented through IoT devices. The number of IoT devices is continuously generating data. This generated data has been passed to the cloud computing layer for processing, but it is very tedious to forward this generated data to the cloud layer for processing due to limited bandwidth and communication latency. Some applications such as health care, military, augmented reality, and online gaming require quick response, but these applications cannot afford communication latency. To prevail over this condition, the fog computing concept has been emerged, which is placed closer to IoT devices. It resides between IoT devices and the cloud computing layer Bonomi et al. (2012). Fog computing widens the capability of cloud computing and resides at the edge of IoT devices. It has better computational power and storage as compared to IoT devices. Thus, it minimizes burdens and traffic to cloud computing and reduces communication latency He et al. (2018). To process real-time data, fog computing contains fog nodes. Fog nodes process data, and less important data can be skipped. Critical data can be used to take a quick decision as well as it can be forwarded to the cloud for storage and further processing. Therefore, data processing can be done at fog rather than at the cloud.

Data generated by IoT devices need to be processed in real time. Therefore, at peak time, some fog nodes may get overloaded and some may be underutilized. In this situation, there is a need for load balancing. Load balancing equally distributes load to all nodes and optimizes resource utilization. If the node gets overloaded, then appropriate load distribution rules must be implemented for the even distribution of load. Due to this diversity of computing capability of node and the dynamic nature of fog nodes, resources may not be fully utilized. Hence, an appropriate load balancing algorithm should be applied to load balancer so that resources should be fully utilized. Thus, it is very essential to design an algorithm that balances the workload in a fog environment.

Various algorithms, such as random, round robin, weighted round robin, least connection and weighted least connection, can be used for load balancing. The random algorithm randomly assigns tasks to any of the available nodes. Round robin assigns time slots to all nodes, and it does not consider the availability of nodes. In the weighted round-robin algorithm, each node is assigned with weight, which is an integer value that denotes node computing capability. However, this algorithm does not consider the size of the task. The least connection algorithm allocates the task to a node that is servicing the least number of requests. The chain failover method circularly arranges node. All requests are sent in the form of a chain. Here, task size and node computing capability should not be considered. In the discussed algorithms, none of the algorithms consider the size of the task and information about nodes. Thus, the proposed algorithm considers only size of task and node information such as computing capability, available memory, and previous history of task completion.

The proposed algorithm is categorized into two phases: load balancing and fault tolerance. In the load balancing phase, our proposed strategy creates a list of valuable nodes and assigns node according to the requirement of the task. In the second phase, our proposed model contains a fault detector that continuously monitors the fog node at the time of task execution for the occurrence of the fault. In this chapter, we proposed fault-tolerant load balancing in a fog computing environment.

Our major contributions are as follows:

- We have a fog computing architecture and system models for the proposed algorithm.
- We have a load balancing and fault-tolerant model, which is suitable to run a proposed algorithm.
- We briefly attempt to described the existing load balancing algorithms.
- We have attempted to evaluate the proposed algorithm with the existing algorithms through simulation using various metrics.

The remaining part of the chapter is structured as follows: Section 4.2 describes background details, while Section 4.3 describes problem statement and definitions. Section 4.4 describes the proposed algorithm, while Section 4.5 describes a study of various load balancing techniques. Section 4.6 discusses performance analysis and evaluation. Section 4.7 concludes the chapter.

4.2 BACKGROUND DETAILS

In this section, the existing fog computing architecture, existing load balancing, and fault-tolerant approaches are described.

4.2.1 EXISTING FOG COMPUTING ARCHITECTURE

Figure 4.1 shows fog computing architecture. It mainly consists of three layers, namely, IoT layer, fog layer, and cloud computing layer.

- IoT Layer (Layer 1): This layer contains smart devices that contain sensors for sensing environment. These sensors generate data, and it is passed to fog layer for processing. Some critical applications require immediate action, and it need to be processed at fog layer for immediate action.
- Fog Computing Layer (Layer 2): This layer is located between the IoT layer and the cloud computing layer. Router, switches, and computing nodes are present in this layer. Data sensed by IoT devices will be forwarded to this layer for processing. Fog nodes may be designed as per requirement, which reside at the edge of IoT devices. Hence, it achieves minimum latency and provides a quick response. Data from IoT devices will be received by the fog controller. The fog controller will be connected to fog nodes, and it will distribute tasks to fog nodes for processing. Generated data from IoT devices processed by fog node, and unnecessary data are skipped, and only useful data is used and passed through cloud server for storage and further processing. This layer reduces data traffic between cloud and IoT devices, thus saving network bandwidth. The following functions are performed by this layer:

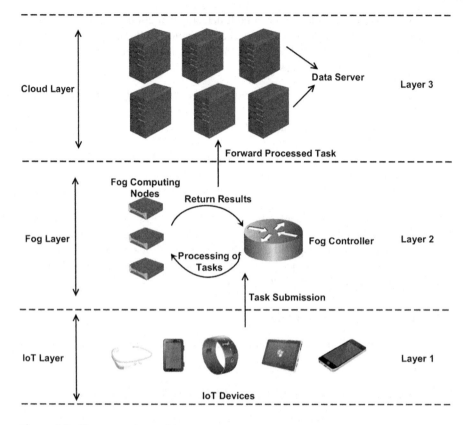

Figure 4.1   Fog computing architecture.

1. Fast processing of data for an immediate action.
2. Data filtering to skip sending unwanted data to the cloud.
3. Collecting data from multiple IoT devices.

- Cloud Computing Layer (Layer 3): This is the topmost layer in the cloud computing environment. It consists of a storage server, which stores data permanently. Processed tasks result which are forwarded to this layer for storage and further computing. It stores data permanently and periodically. The following functions can be enumerated by this layer:

  1. Taking data backup at a periodic interval;
  2. Taking an incremental update of data on the data server and store data permanently;
  3. Processing queries and retrieving data from the data server;
  4. Performing a detailed analysis of data;
  5. Collecting data from multiple IoT devices.

4.2.2   EXISTING LOAD BALANCING STRATEGIES

Chien et al. (2018) introduced a software-defined network for load balancing in IoT applications. This method uses the combination of a software-defined network (SDN) and service function chaining (SFC) for load balancing. First of all, it classifies various types of services required by IoT devices and then prioritizes the services. Then, it uses a heuristic algorithm to reduce the load of each service function. Xu et al. (2018) proposed a load balancing technique in fog computing using dynamic resource allocation. The proposed method achieves load balancing for all types of computing nodes in the fog and the cloud. Manju et al. (2019) proposed a load balancing strategy for fog computing. Fog computing environment contains end user, network resources, and cloud resources. In the proposed strategy, network resources act as a central node that evenly distributes load among fog nodes and reduces response time of task processing. Ningning et al. (2016) proposed a graph partitioning method for load balancing in a fog computing environment. The proposed method uses cloud atomization for load balancing in a fog computing environment.

Workload balancing in a fog computing environment has been discussed by Verma et al. (2018). This chapter introduces a load balancing strategy for a fog-cloud-based environment. Data is replicated in the proposed algorithm for maintaining data in fog. However, in the article by Kapsalis et al. (2017), the tasks are portrayed by their computational need and are allotted to the proper host. An energy-aware load balancing and scheduling (ELBS) method based on fog computing has been proposed by Wan et al. (2018). Gamal et al. (2019) uses an osmosis theory from chemistry for load balancing in cloud computing. The author combines the osmotic hybrid artificial bee and ant colony optimization (OH-BAC) load balancing algorithm. The proposed algorithm OH-BAC decreases energy consumption, virtual machine migrations, and the number of shutdown hosts. Tang et al. (2018) proposed dynamical load-balanced scheduling (DLBS) approach for big data centers in cloud computing. The proposed algorithm uses a load imbalance factor that adapts dynamical network states and traffic requirements. Xu et al. (2013) proposed load balancing in the cloud, which is based on cloud partition. It uses game theory for efficient load balancing. In this article, the cloud is divided into three categories such as idle, normal, and overloaded based on the value load degree. If the value of load degree is zero, then the cloud is idle; if the value of load degree is between zero and the highest, then the cloud load is normal else cloud status will be overloaded.

4.2.3   EXISTING FAULT TOLERANCE STRATEGIES

Numerous articles have been published on fault tolerance aspect in grid and cloud computing, but very few research articles have been published on the fault tolerance aspect in fog computing. Wang et al. (2020) proposed a fault-tolerant data-processing approach in health-care application. It uses reduced variable neighborhood search (RVNS), which ensures fault-tolerant data transmission between storage and processing node. The proposed approach uses directed diffusion and limited flooding for reliable data transmission. The proposed approach improves successfully delivered ratio and optimizes resource allocation. Enokido et al. (2019) proposed a fault-tolerant tree-based model for fog computing environment. In this model, fog nodes are arranged hierarchically for processing. Egwutuoha et al. (2012) uses a process-level redundancy approach for fault tolerance in the cloud computing environment. The proposed approach reduces wall clock time for the execution of computational applications. Egwutuoha et al. (2013) presented energy-efficient fault tolerance for high-performance computing in the cloud environment. The proposed technique uses process-level migration among computing nodes, and it does not use the spare node or redundant technique. Hasan et al. (2019) proposed flexible fault tolerance framework (FFTF) in the cloud, which categorizes tasks into various categories to implement the level of fault tolerance. Ding et al. (2017) proposed a fault-tolerant elastic scheduling algorithm, which also provides dynamic resource provisioning by using the resource migration technique. The algorithm achieves fault tolerance and high resource utilization for workflow in cloud systems.

From this literature survey, it is observed that various techniques have been used for load balancing and fault tolerance in cloud and fog computing. However, none of the articles consider fault tolerance along with load balancing in fog computing. Hence, fault-tolerant load balancing algorithm has been proposed for fog computing.

4.3 PROBLEM STATEMENT, NOTATIONS, AND DEFINITIONS

4.3.1 PROBLEM STATEMENT

Fog nodes execute tasks on fog nodes when they receive tasks from IoT devices. However, in general, any available host is randomly assigned to tasks for execution. However, if the resource requirement of task is greater than the fog node resource, then the fog computing node cannot execute the task. If resource requirement is lesser than or equal to the fog node resource, then the task can be executed; if more tasks are needed, then some fog nodes may be overloaded. As a result, the quality of service can be declined, and the waiting time of task may increase. At the same time, during task execution, some fog nodes, may fail. Hence, there is a need for optimal load balancing and fault-tolerant strategy in a fog environment.

This problem statement is formulated as follows. Let F denote set of fog nodes available at the fog computing center which is to be deployed in fog computing for the execution of a set of tasks t. Define a function X_m whose value is 1 if fog node f is assigned to task t, 0 otherwise.

$$X_m = \begin{cases} 1, & \text{if task t deployed on fog node f} \\ 0, & \text{otherwise} \end{cases} \tag{4.1}$$

The assignment of the task to fog node is said to be efficient if makespan, flowtime, and average execution time are minimized and successful execution rate maximized in a fog computing environment.

Table 4.1
Notations

Notation	Meaning
F	Set of fog node
C_i	Computing capability of fog node i
CT	Sum of completion time of all task
f_i	i^{th} fog node
ST_i	Starting time of ith task
ET_i	Expected execution time of ith task
T	Total number of task
$N_{(succ)}$	Number of tasks successfully executed within deadline

4.3.2 NOTATIONS AND DEFINITIONS

Here, notations used in the article are defined followed by a definition that is used in the proposed algorithm which is shown in Table 4.1.

4.4 PROPOSED MODEL FOR LOAD BALANCING AND FAULT TOLERANCE

In this section, a proposed model is described which is suitable to run the proposed algorithm. Let n be the tasks number that need to be processed on fog nodes within time δt which is to be executed on f fog nodes. Let at time delta t, m fog nodes are available for task execution. This problem is described as finding the best fog node for task execution. It is defined as $P = <T_i F>$, where T_i represents a set of available tasks and F represents a set of available fog nodes. Fog node is described as $X = <f_{(CPU)}, f_{(mem)}, \text{prev-hist}>$, where $f_{(CPU)}$ represents CPU available at fog node, $f_{(mem)}$ represents memory available at fog node and prev-hist indicates the previous history of task execution. The goal is to execute the task within time δt on the fog node efficiently so that a quick response can be obtained. Figure 4.2 shows the proposed model for algorithm.

4.4.1 FOG COMPUTING ENVIRONMENT

Fog computing consists of a set of nodes f_1, f_1, f_n connected through communication links, and each fog node has computing capability. Fog controller is a software that runs in the fog computing environment. Fog controller manages fog nodes and receives a task from IoT devices. In building a fog environment, the system environment should follow the following conditions:

- Each fog node that joins in the fog computing environment has to provide the computing capability and hardware information;

Table 4.2
Definitions

Sr. No.	Definition
1	**Computing Capability of Fog Nodes:** It is defined as summation of computing capability of all fog nodes.

$$C = \sum_{i=1}^{n} C_i \qquad (4.2)$$

| 2 | **Completion Time:** Completion time is sum of start time and execution time of task. |

$$CT_i = ST_i + ET_i \qquad (4.3)$$

| 3 | **Flowtime:** Flow time is sum of completion time of task Chekuri et al. (2004). |

$$Flowtime = \sum_{n=1}^{n} CT_i \qquad (4.4)$$

| 4 | **Average Execution Time:** It is described as completion time of all tasks divided by number of task Mao et al. (2016). |

$$Avg\ Execution\ Time = \sum_{i=1}^{n} CT_i / T \qquad (4.5)$$

| 5 | **Makespan:** It is computed as the maximum of completion time(CT) Raju et al. (2013). |

$$Makespan = max(CT_i) \qquad (4.6)$$

| 6 | **Success Execution Rate:** If completion time of task is less than its own deadline then it is said to be successfully executed Yunmeng et al. (2019). |

$$Success\ Execution\ Rate = N_{(succ)} / T * 100\% \qquad (4.7)$$

- Fog controller has full control over the fog computing environment and monitors the state of each fog node;
- Contribution of each fog node has been recorded, which can be used for making decisions in the future.

In this environment, the fog node has an important role for task execution. If any node is idle and is in state of providing a resource for the execution of tasks, a 'free' message with computing capability will be sent to the fog controller. When the fog node is not in the state for execution of tasks, 'busy' message will be transmitted to the fog controller.

4.4.2   COMPONENTS OF PROPOSED MODEL

The components of proposed model are described as follows:

- **Fog Controller:** Fog controller executes and manages the task in the fog computing environment. Arrived tasks are placed in the task queue, and then the fog controller assigns these tasks to fog nodes. The fog controller collects complete information of each fog node using CPU utilization, memory capacity, and state of each node. As shown in Figure 4.2, as soon as the fog controller receives tasks from IoT devices, the fog controller collects fog node information based on task requirement and sends a list to the decision maker of load balancing.
- **Load Balancing Decision Maker:** The decision maker will select the node for the execution of tasks.
- **Task Dispatcher:** It assigns tasks to fog nodes.
- **Fault Detector:** The fault detector keeps a track of the state of the fog node, and if fog nodes fail, it sends a fail message to the fault monitor.
- **Fault Monitor:** If the fault monitor receives a fail message, it again reschedules the task and forwards it to the fog controller.

Fog nodes are dynamic, and their state can change anytime. Hence, while selecting a fog node for task execution, CPU utilization is considered. Therefore, concept of

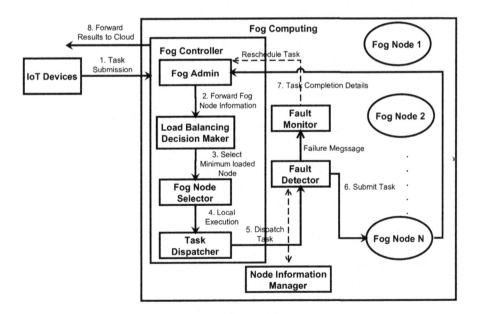

Figure 4.2   Proposed load balancing and fault tolerant model.

valuable fog node (VFN) has been proposed for selecting a fog node for task execution. Fog controller uses a threshold value for selecting VFN. The threshold value is calculated as the difference between the load on the fog node and the resource required for the execution of the task. The node which is within the threshold value is regarded as an effective fog node. Fog controller prepares a list of VFNs in ascending order. Since the state of the fog node can be changed anytime, so it is very important to update the list of effective fog node. However, if it is changing very frequently, then it will be a burden to the fog controller. Therefore, two approaches can be considered for calculating VFN:

- Stable Fog Environment: If the state of fog node does not change frequently, then the list of valuable fog nodes is updated only when any new node joins or leaves fog computing environment.
- Unstable Fog Environment: If the state of the fog node changes frequently, then updating the list of VFNs adds burden to the fog controller. Therefore, the VFN list is updated only when the node value reaches a threshold value or node leaves a fog environment.

4.4.3   ALGORITHM DESCRIPTION

In this section, details of the proposed algorithm are discussed. The purpose of this algorithm is to assign tasks optimally to fog nodes. For the proposed algorithm, a set of tasks is given as input. To assign tasks, fog nodes are required. Tasks are executed only on VFN, and VFN is considered only when the node reaches a threshold value. The available CPU computing capability, available memory space, and previous history of task execution are considered for finding valuable fog node. Task execution gets fail notification only when the fog node fails in execution due to hardware failure. Since tasks are independent, they can be executed on any of the effective nodes. The node with the highest VFN value is considered an effective node. However, in a fog environment, due to the dynamic nature of fog nodes, VFN value may change frequently. Thus, tasks should be dynamically allocated to nodes. There can be two variations: first, when the fog controller comes to know that node is unavailable; second, fog controller receives a message that the fog node cannot provide resources. When any of the situations occur, the fog controller selects the next VFN. If any node is not able to complete the assigned task, then the node with the highest VFN will be selected from the list of effective nodes. The fog controller collects status information from all nodes, and if nodes are busy, then no task is submitted. After the forwarding task to the fog node, if the fog controller receives heavy load status, then the task will be transferred to a less loaded node. In exchange load from heavily loaded node to lighter node, the fog controller should pay attention such that lightly loaded node should not get overloaded. Whenever the node is allocated for task execution, the fault detector continuously monitors the status of the fog node, and if the node fails, then it sends a fail message to the fault monitor. The fault monitor then rescheduled the task and sends it to the fog controller. Pseudocode 1 describes the proposed algorithm.

4.4.3.1 Pseudocode of FTLB Algorithm

Algorithm 1 Pseudo code of Proposed FTLB Algorithm

Input: A set of tasks and Nodes
Output: Completion Time of Tasks
1. **For** all submitted task in the queue **do**
2. **While** there are no unscheduled task
3. Sort the list of valuable fog nodes in ascending order
4. Select the first fog node N_j
5. Schedule task T_i on Node N_j
6. Update load on Node N_j
7. Fault detector monitor node status while task execution
8. **If** Node fails while execution
9. Reschedule the task
10. **Else**
11. Delete task T_i from unscheduled tasks
12. **End if**
13. **End if**
11. repeat steps 3 to 11 until all task are mapped
13. **End While**

4.5 A BRIEF DESCRIPTION OF LOAD BALANCING ALGORITHMS

4.5.1 ROUND ROBIN

It is a simple strategy for load balancing among computing nodes. Here, all computing nodes are configured to provide the same set of services. In this strategy, all tasks have been assigned with fixed time slots and are executed cyclically. Tasks that are in queue are for execution, the node has been assigned to them for a certain time quantum. If execution gets completed in a certain time quantum, then task execution gets completed else task will be put in queue Calheiros et al. (2009). It does not consider task size and availability of the resource. As shown in Figure 4.3, there are three computing nodes behind a load balancer. The first task assigned to the first computing node, the second task assigned to the second computing node, and the third task assigned to the third computing node. Since the third node is the last, the fourth task will be forwarded to the first node, the fifth task will be forwarded to the second node, and so on, which is cyclical. However, if node one has more computing power and memory as compared to nodes two and three, then it should handle more tasks as compared to other nodes. Load balancer running on a round-robin algorithm can differentiate nodes. Even though nodes have different computing capabilities, the load balancer will distribute tasks equally to all nodes. As a result, there will be slow processing of tasks on nodes two and three.

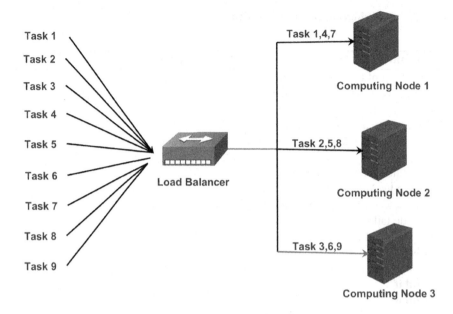

Figure 4.3   Round robin for load balancing.

4.5.2 WEIGHTED ROUND ROBIN

A weighted round-robin algorithm is used when computing nodes have different computing capabilities. Each node is assigned with weight, which is an integer value that denotes node computing capability. The computing node with the highest value will be selected for task execution Alnowiser et al. (2014). In this algorithm, some nodes have a higher specification than other nodes. As shown in Figure 4.4 node one has higher specification than nodes two and three. Here, node has 2X more computing capabilities than nodes two and three. Therefore, tasks 1,2,3,4, and 5 are assigned to node one, and the remaining tasks are assigned to nodes two and three.

4.5.3 RANDOM ALGORITHM

The random algorithm assigns tasks to nodes by random using a random number generator. If a load balancer receives a large number of tasks, this algorithm will distribute tasks to nodes evenly Yang et al. (2003). As shown in Figure 4.5, tasks are randomly distributed to any of the nodes.

4.6 EXPERIMENTS

4.6.1 SIMULATION ENVIRONMENT

The simulation parameters are shown in Table 4.3. To analyze the effectiveness, the number of tasks changes from 100 to 500. Simulation parameters are given in

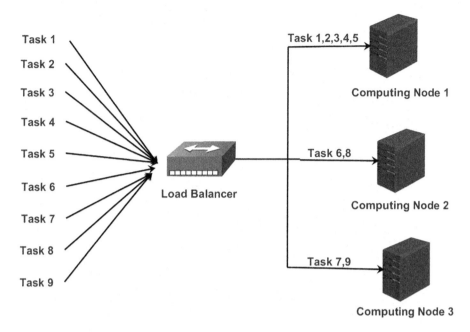

Figure 4.4 Weighted round robin for load balancing.

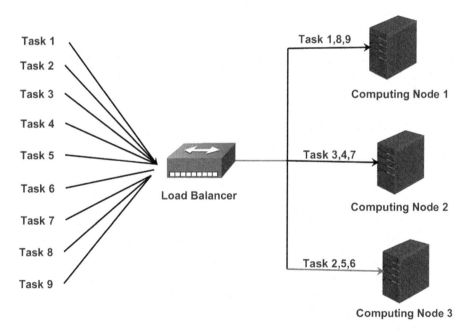

Figure 4.5 Random algorithm for load balancing.

Table 4.3

Machine Configuration

Model	Intel PC
Processor	2.oo GHz Intel(R) Core(TM) i3-5005U
Operating system	Windows10
RAM	4 GB

Table 4.4

Simulation Configuration

Simulation Tool	CloudSim
Task ETC	1–10 (Random distribution)
Task arrival	Poisson distribution
RAM	512 MB
MIPS	1000
Simulation starts at	100 tasks
Simulation end at	500 tasks

Table 4.4. A systematic evaluation has been carried out to evaluate the performance of the proposed load balancing strategy. We compared the proposed algorithm with random, round robin, and weighted round robin. The performance evaluation has been done using four metrics: makespan, flowtime, average execution time, and success execution rate.

4.6.2   PERFORMANCE ANALYSIS

In this section, a comparison of the proposed algorithm with the existing algorithm using makespan, flowtime, average execution time, and success execution rate is made.

Makespan Comparison: The proposed algorithm is compared with RR, WRR, and random algorithms based on makespan. Figures 4.6 and 4.7 show the performance of algorithms for 5 and 10 fog nodes, respectively. With the increase in the number of tasks, the makespan value of all algorithms increases. When the number of tasks is 100, there is not much difference between makespan values. The reason is that there are fewer tasks at the initial stage, and hence the time required for processing tasks is less. However, as the number of tasks increases, makespan values of all algorithms increase. The makespan value is lesser than the other three algorithms. The reason is that random algorithms randomly assign tasks to any of the available nodes irrespective of computing capability. This will increase the processing time of tasks. In the round robin, tasks are cyclically assigned for fixed time quantum, and

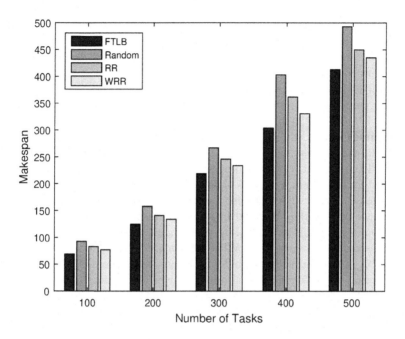

Figure 4.6 Makespan for 5 node.

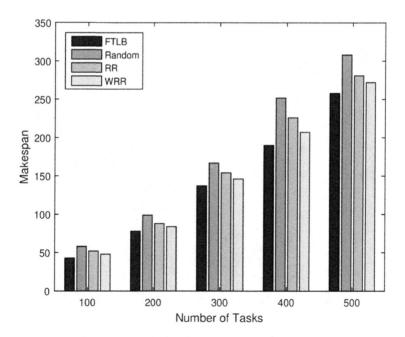

Figure 4.7 Makespan for 10 node.

if the task is not completed within a certain time quantum, then it has to wait, which will increase processing time. WRR algorithm assigns weight to node, and tasks are assigned accordingly to the node. However, it does not consider the task size. However, the proposed FTLB algorithm considers CPU utilization to assign a task to the optimal node. From Figure 4.6 and 4.7, we can analyze that the proposed FTLB algorithm gives better performance as compared to other load balancing algorithms (random, RR, and WRR) for makespan.

Flowtime Comparison: Here, the flowtime proposed algorithm is compared with RR, WRR, and random algorithms. Flowtime is defined as the sum of completion time of all tasks, and it is calculated by formula(4). Through simulation, flowtime is calculated for a different set of tasks after execution on fog nodes. If the number of tasks increases, then the flowtime also increases. Figures 4.8 and 4.9 demonstrate the performance of algorithm for 5 and 10 fog nodes, respectively. The performance of the random algorithm is worst, and the performance of WRR lies between RR and the proposed algorithms. Initially, WRR performance is good, but with an increase in a number of tasks, its adaptive ability becomes weaker. From Figures 4.8 and 4.9, by comparing the performance of all four deployed algorithms, it is concluded that the FTLB algorithm gives less flowtime as compared to existing algorithms.

Average Execution Time Comparison: Here, the average execution time of the proposed FTLB algorithm is compared with RR, WRR, and random algorithms. Average execution time is calculated by changing a number of tasks as shown in Figure 4.10 and 4.11. As the number of tasks increases, the average execution time gradually

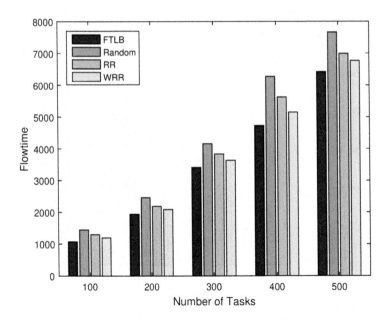

Figure 4.8　Flowtime for 5 node.

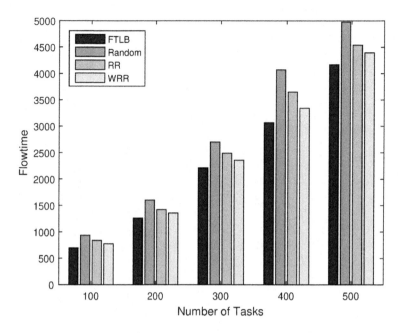

Figure 4.9   Flowtime for 10 node.

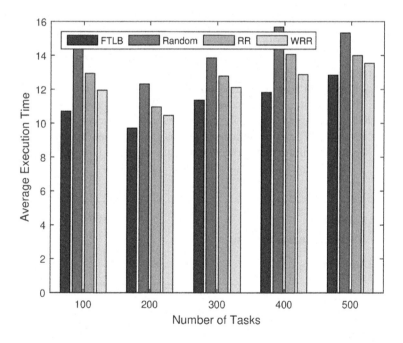

Figure 4.10   Average execution time for 5 node.

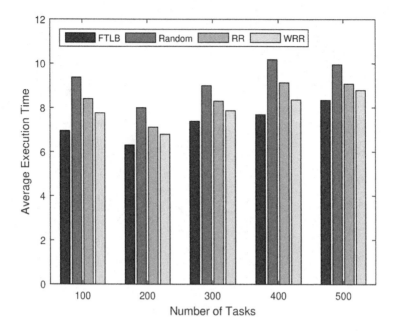

Figure 4.11   Average execution time for 10 node.

increases. The proposed FTLB algorithm assigning the task to the best possible node, thus reducing the execution time of the task. From Figures 4.10 and 4.11, it is observed that the proposed FTLB algorithm gives less values of average execution time as compared to existing algorithms.

Success Execution Rate Comparison: Here, the success execution rate of the proposed FTLB algorithm is compared with RR, WRR, and random algorithms. It is calculated by formula(7). Table 4.5 shows the success execution rate generated by four algorithms. When the number of the task is 100, then all algorithms have the same success execution rate. However, with the increase in the number of tasks,

Table 4.5
Success Execution Rate

Algorithm	Tasks				
	100	**200**	**300**	**400**	**500**
FTLB	1	0.95	0.92	0.89	0.84
RR	1	0.90	0.86	0.80	0.78
WRR	1	0.90	0.91	0.82	0.80
Random	1	0.85	0.82	0.75	0.70

the success rate decreases. The success rate of FTLB and WRR slowly decreases, while the random algorithm has the lowest success execution rate. It is observed that a better efficient assignment of tasks better is the success execution rate. Simulation results demonstrate that the FTLB algorithm has a higher success execution rate as compared to the other three algorithms.

The following are the reasons for better performance of the proposed FTLB algorithm:

1. A strategy has been proposed that finds the best possible fog node for task execution. The best fog node is calculated by using CPU availability, memory availability, and previous history of task execution;
2. Existing load balancing algorithms do not consider task size and node computing capability. However, our proposed algorithm considered these factors;
3. Proposed algorithm finds the best node for task execution.

4.7 CONCLUSION

At peak time, some fog computing nodes are overloaded while other are underloaded. Load should be equally distributed among fog nodes for efficient resource utilization. During task execution, fog node may fail, hence continuous monitoring is required on fog nodes. Hence, an algorithm for handling fault tolerance and load balancing has been proposed, which performs load balancing and fault tolerance in a fog computing environment. The proposed algorithm finds valuable fog nodes, and the task has been mapped to the best possible fog node. The proposed FTLB algorithm also finds fault in node during task execution, and if the node fails, then the task will be rescheduled. This algorithm helps in efficient resource utilization and fault tolerance. To evaluate the proposed FTLB algorithm, several simulations were carried out and compared to FTLB, RR, WRR, and random algorithms based on makespan. Simulation results showed that FTLB has the smallest makespan value. A comparison based on flowtime has been carried out, and again FTLB had lower value of flowtime as compared to RR, WRR, and random algorithms. Then, FTLB has a less average execution time of tasks as compared to other algorithms. Finally, the success execution rate of all four algorithms had been calculated under the same conditions, and FTLB showed a better success execution rate as compared to the other three algorithms. Based on four comparisons, it is concluded that FTLB is an efficient implementing load balancing method.

REFERENCES

Alnowiser A, Aldhahri E, Alahmadi A, Zhu MM. (2014) Enhanced Weighted Round Robin (EWRR) with DVFS Technology in Cloud Energy-Aware. *In: 2014 International Conference on Computational Science and Computational Intelligence*, vol. 1,320–326.
Bonomi F, Milito R. (2012) Fog Computing and its Role in the Internet of Things. *Proceedings of the MCC workshop on Mobile Cloud Computing*

Calheiros R, Ranjan R, De Rose C, Buyya R. (2009) CloudSim: A Novel Framework for Modeling and Simulation of Cloud Computing Infrastructures and Services.

Chekuri C, Goel A, Khanna S, Kumar A. (2004) Multi-processor Scheduling to Minimize Flow Time with Resource Augmentation, *In: Proceedings of the Thirty-sixth Annual ACM Symposium on Theory of Computing STOC '04*, 363–372.

Chien WC, Lai CF, Cho HH. (2018) A SDN-SFC-based service-oriented load balancing for the IoT applications. *Journal of Network and Computer Applications*, 114.

Ding Y, Yao G, Hao K.(2017) Fault-tolerant elastic scheduling algorithm for workflow in Cloud systems. *Information Sciences*, 393, 47–65.

Egwutuoha IP, Chen S, Levy D, Selic B.(2012) A Fault Tolerance Framework for High Performance Computing in Cloud. *In: 2012 12th IEEE/ACM International Symposium on Cluster, Cloud and Grid Computing (ccgrid 2012)*, 2012, 709–710.

Egwutuoha IP, Cheny S, Levy D, Selic B, Calvo R.(2013) Energy Efficient Fault Tolerance for High Performance Computing (HPC) in the Cloud, *In: 2013 IEEE Sixth International Conference on Cloud Computing*, 2013, 762–769.

Enokido T, Takizawa M, Nakamura S, Duolikun D, Oma R.(2019) A Fault-Tolerant Tree-based fog computing model, *International Journal of Web and Grid Services*, 15, 219–224.

Gamal M, Rizk R, Mahdi H, Elnaghi BE. (2019) Osmotic Bio-Inspired Load Balancing Algorithm in Cloud Computing. *IEEE Access*, 7, 42735–42744.

Hasan M, Goraya MS. (2019) Flexible fault tolerance in cloud through replicated cooperative resource group, *Computer Communications*, 145, 176–192.

He D, Qiao Y, Chan S, Guizani N.(2018) Flight Security and Safety of Drones in Airborne Fog Computing Systems. *IEEE Communications Magazine*, 56(5), 66–71.

Kapsalis A, Kasnesis P, Venieris IS, Kaklamani DI, Patrikakis CZ.(2017) A Cooperative Fog Approach for Effective Workload Balancing. *IEEE Cloud Computing*, 4(2), 36–45.

Manju AB, Sumathy S. (2019) Efficient Load Balancing Algorithm for Task *Preprocessing in Fog Computing Environment* In: Satapathy SC, Bhateja V, Das S, editors. Smart Intelligent Computing and Applications. Singapore: Springer Singapore; 291–298.

Mao Y, Zhang J, Letaief KB. Dynamic Computation Offloading for Mobile-Edge Computing With Energy Harvesting Devices, *IEEE Journal on Selected Areas in Communications*, 34(12), 3590–3605.

Ningning S, Chao G, Xingshuo A, Qiang Z.(2016) Fog computing dynamic load balancing mechanism based on graph repartitioning. *China Communications*, 13(3), 156–164.

Raju R, Babukarthik RG, Chandramohan D, Dhavachelvan P, Vengattaraman T. (2013) Minimizing the Makespan using Hybrid algorithm for cloud computing.*In: 2013 3rd IEEE International Advance Computing Conference (IACC)*, 957–962.

Tang F, Yang LT, Tang C, Li J, Guo M. (2018) A Dynamical and Load-Balanced Flow Scheduling Approach for Big Data Centers in Clouds, *IEEE Transactions on Cloud Computing*, 6(4), 915–928.

Verma S, Yadav AK, Motwani D, Raw RS, Singh HK.(2018) An efficient data replication and load balancing technique for fog computing environment. *In: 2016 3rd International Conference on Computing for Sustainable Global Development (INDIACom,* 2888–2895.

Wan J, Chen B, Wang S, Xia M, Li D, Liu C.(2018) Fog Computing for Energy-aware Load Balancing and Scheduling in Smart Factory. *IEEE Transactions on Industrial Informatics*, 1–1.

Wang K, Shao Y, Xie L, Wu J, Guo S.(2020) Adaptive and Fault-Tolerant Data Processing in Healthcare IoT Based on Fog Computing. *IEEE Transactions on Network Science and Engineering*, 2020, 7(1), 263–273.

Xu G, Pang J, Fu X. (2013) A load balancing model based on cloud partitioning for the public cloud, *Tsinghua Science and Technology*, 18(1), 34–39.

Xu X, Fu S, Cai Q, Tian W, Liu W, Dou W (2018) Dynamic Resource Allocation for Load Balancing in Fog Environment. *Wireless Communications and Mobile Computing*, 1–15.

Yang K, Guo X, Galis A, Yang B, Dayou L. (2003) Towards Efficient Resource on-Demand in Grid Computing. Operating Systems Review, 37, 37–43.

Yunmeng D, Xu G, Ding Y, Meng X, Zhao J. (2019) A 'Joint-me' Task Deployment Strategy for Load Balancing in Edge Computing, *IEEE Access*, 1–1.

5 Real-Time Solar Energy Monitoring Using Internet of Things

V. Vijaya Rama Raju and J. Praveen
GRIET
Hyderabad, India

CONTENTS

5.1 Introduction 87
5.2 Existing Literature Based on IoT 91
5.3 Application of IoT to Monitor the Solar Energy Resources...........................95
     5.3.1 Major Advantages of IoT in Renewable Energy Applications...........95
     5.3.2 IoT Device for Data Acquisition – A Case Study 96
     5.3.3 Procedure to Acquire Data from the Plant using IoT Sensor Board...96
     5.3.4 System Architecture...97
5.4 Analysis of Results 97
5.5 Conclusion 102
5.6 Future Scope 102
References...102

5.1 INTRODUCTION

Electricity has become indispensable in our daily life. In every walk of human life, electricity is being used for various applications such as transport, illumination, cooking, freezing, cooling, and several other applications. Most of the appliances in residential, commercial, and industry sectors are run on electricity. The usage of electricity usage keeps on increasing with a positive gradient every day. To meet the exponentially raising demand, a parallel amplification of energy generation needs to be done as demand will also inflate with the increase of population. This could be observed by looking at the predictions about the number of units required to meet the exponentially raising demand and growth rate of 8% and 7%, respectively, for the upcoming 5-year plans in India alone as shown in Figures 5.1 and 5.2.

The majority of electrical energy is produced using three physical phenomena, namely, electromechanical, chemical, or photovoltaic effects. Electromechanical effect was the most popular among the three, and nearly 70% of the electrical needs

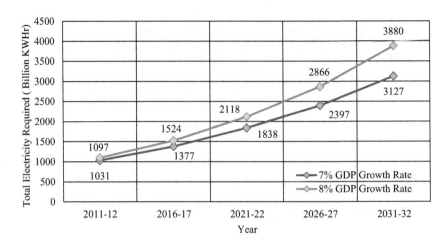

Figure 5.1   Electrical energy prediction in terms of billion kWh till 2031 in India.

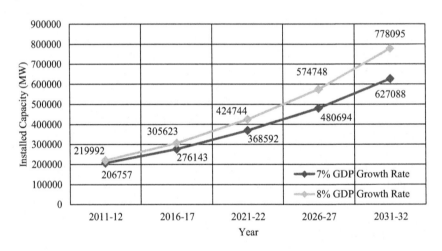

Figure 5.2   Expected installed capacity in terms of MW till 2031–2032 in India.

were met through this process only. Faraday's law of electromagnetic induction of converting mechanical energy into electrical energy was the basic principle behind the energy generation. Chemical effect is used for producing electrical energy after the invention of battery by Alessandro Volta. These are also used to store the energy in chemical form and can be converted to electrical energy on need basis. The third type of source for electricity generation is photovoltaic cell that converts light energy into electrical energy. This is known as photovoltaic effect, which is used in solar power plants. Electrical energy generated from these power plants is renewable.

Solar energy is an effective renewable resource that can become a cost-efficient and eco-friendly alternative for fossil fuels. The energy that emanates for 90 min on a usual sunny day would meet a year's electric requirements of the entire earth. It also thoroughly avoids hazardous greenhouse gas (GHG) emissions and other pollutants during the operation of solar PV systems. Millions of solar panels are getting installed across the globe every day to stabilise the grid. Around 40% of the renewable energy is getting generated from solar PV systems.

In the initial days of solar PV cell invention, power generated from conventional energy resources is much cheaper than the energy generated by using solar PV panels due to the cost factor of panels. Especially in rural areas where there is no electric connectivity, off-grid solar system is a very good solution for the residential consumers or farmers. As the conversion technology is getting advanced day by day, the cost of the panels is getting reduced making the solar power plants a better solution to meet the energy demands. Even the federal systems are also encouraging the solar installations in terms of subsidies and set targets in their strategic plans. This has dramatically increased the deployment of solar power plants across the country. Changes in the climatic conditions force the nations to shut down the generating stations that use fossil fuels to reduce their carbon footprint. Domestic customers who were consumers in earlier days have now become prosumers by installing solar panels on their rooftops. In the sense that they produce the energy during the daytime and pump the excess energy into the grid for consuming the same during the night time leading to net zero.

To generate electrical energy in bulk for consumers like commercial/business establishment numerous panels are integrated in arrays that is considered as solar power plant as shown in Figure 5.3.

Energy generated from the solar panel depends on the radiation incident on the panels that get converted into electrical energy. In case of reduction in irradiance on the panels due to shadow or dust accumulation, generation gets affected. Continuous monitoring of the solar power plant is essential to enhance the performance of

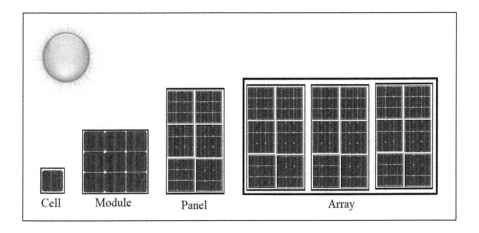

Figure 5.3 Cell – Module – Panel – Array structures.

the system. Nowadays, electric power determines standard of living. Energy consumption in daily life increases, while other energy sources are rapidly depleting. To meet the growing demand for energy, one has to explore the resources that are renewable and eco-friendly. These resources are broadly classified into two categories: non-renewable and renewable energy resources. Non-renewable resources, such as such as coal, natural gas, and nuclear fuel, are exhausting and can't be renewed in short time. While renewable sources, such as solar, geothermal, wind and tidal energy, are inexhaustible and can be reused. Hence, these are called sustainable technologies.

As per the International Energy Agency (IEA) report (Renewables, 2019), the booming source of energy will be renewable energy resource. Especially solar and wind are the major sources which are in the forefront, being matured in terms of technology and affordability. Continued increase in demand for energy across the globe mandates to leverage the advantages of these resources that reduce on the adverse impact on environment due to the use of fossil fuels.

The Internet of Things (IoT) is a system made up of digital devices with high communication and computing facilities, mechanical machines with digital interface, numerous sensors that can transfer huge amount of data over communication channels without human intervention. Sensors are connected to the devices for the production, transmission, and distribution of IoT applications in the production of renewable energy. These devices allow businesses to monitor and control devices in real time. This leads to lower maintenance costs and reduced dependency on the fossil fuels that are limited. The use of renewable energy sources already offers many advantages over conventional sources. The introduction of the IoT will allow to make more use of these pure energy resources. Studies conducted by researchers in areas like dust accumulation have proved that efficiency has been reduced by 50% due to improper maintenance of solar panels and the system.

Intrusion of IoT into electric grid will benefit the consumers. This also allows the integration of renewable energy resources and electric vehicles into the electric grid. Incorporating the sensors across the transmission and distribution networks in a large way, system operator will get real-time picture of the system at a glance. This will help the operator to take necessary actions at a right time. Even the sensors will also give alarms in case of any abnormalities in the system. The data from the sensors can be used in different forms like forecasting, real-time applications such as voltage regulation, frequency control, reactive power management, and so on. IoT can automate some of these decisions. Automated systems work more efficiently than human-dependent systems. In the event of a shutdown, the smart switch will automatically isolate the problem area. IoT devices can divert power and turn lights ON quickly. All these actions will improve the system stability in a big way.

This remaining part of the chapter is organised as follows. Status of IoT-based technology for solar energy monitoring is discussed in Section 5.2. IoT devices used to develop for monitoring the solar systems and their advantages are elaborated in Section 5.3. Section 5.4 discusses a recorded data from the IoT devices and their applications. The summary of the IoT-based solar system is concluded in Section 5.5. Section 5.6 presents insights about IoT-based solar energy system.

5.2   EXISTING LITERATURE BASED ON IoT

Solar energy is becoming more popular in the world due to its infinite availability and eco-friendly factor. However, with the development of changing technology, solar energy production has become much cheaper. The photovoltaic solar system produces a lot of energy, necessitating real-time monitoring of the performance of such systems. This requires monitoring of various parameters using IoT. The basic purpose of these devices is to communicate between the field equipment/sensors and the cloud. Also, the devices can be used to collect the data from cloud and send the information to the field equipment wirelessly.

Observing parameters on continuous basis is utmost important in solar energy production systems. Information regarding solar photovoltaic voltage system's voltage, current, and temperature can be gathered with real-time monitoring using various sensors. Remote monitoring of solar energy systems causes a lot of problems because it must be solved by dealing with electrical energy, current, power, and panel temperature and synchronising the actual data of the solar system from time to time. To gather the data from the field, microcontrollers like Arduino can be used to send the data to cloud.

Monitoring systems using IoT are applicable for both rooftop systems and ground-mounted solar systems. The usage of IoT solves many of the concerns that are normally encountered in remote monitoring of PV panels with a very little amount of effort and investment. This could be achieved by eliminating the integration of complicated hardware that is used for monitoring and rendering physical efforts for data collection. IoT allows the organisation to seamlessly access the real-time data in one's own cloud infrastructure and integrate with the existing Information Technology systems in the system. This allows to extract the data in a more secured environment from the remote assets in the organisation using cloud platform.

Advantages of Internet services can be leveraged to control the equipment which is at remote locations. Smart world can only be achieved only with devices and processes that are smarter. These IoT devices are vital in the implementation of smart cities. Advancements in information and communications technology (ICT) have all the capabilities to transform the system. This is evident from the deployment of Wi-Fi and 4G-LTE wireless Internet access that are ubiquitous. Major areas at which the IoT technology finds its applications in renewable energy resources are the sensors that are connected with generators like solar PV panels or wind turbines in case of wind energy and the power transfer equipment. IoT devices augment the utilities by providing monitoring and control functions to enhance the performance of the plants from remote locations to utilise power optimally and also for identification of faults intelligently (Andreas S. Spanias, 2017). This will definitely reduce operational costs of the utilities, and more importantly, this will reduce our dependency on the fossil fuels that are nonrenewable and scarce. Renewable energy sources have several advantages over the nonrenewable energy resources such as availability in a long run, less maintenance, low cost, distributed, and so on. Application of IoT technology will definitely help us to extract maximum out of these renewable resources that are clean and inexhaustible.

Solar power plant consists of thousands of solar panels, each of which independently generates a certain amount of electricity. Solar inverter panels are connected to each other at a focal point where the solar string inverter that converts the DC to AC energy, which can be used to power electrical components. Most systems use a network of wired sensors to monitor various aspects of the process. The success of monitoring system depends on the efficient and reliable communication network. For any reason, the network crashes, devices will not be able to send data, which could result in the loss of important data that indicate performance issues or misinterpretation of other available data.

In the present day, almost all surveillance systems have wire sensors/components. When efficient, cost and maintenance costs control the number of sensors that can be delivered in a given system. Hardware and tracking costs are dramatically reduced with the wireless IoT configuration, resulting in greater data collection, tracking capabilities, and more practical/AI applications. Many systems are utilising cellular networks or Wi-Fi for communication. Cellular coverage is popular, and its geographical scope is vast, but it is very expensive. Wi-Fi presents its own challenges as many IT departments avoid allowing third-party IoT devices on their networks for security reasons. Solar power systems for commercial use often come as part of a package with tracking services and/or cellular customer components. An intermediate choice between cellular and Wi-Fi is IoT's exclusive network. Specially designed to connect to the Internet, it works like Wi-Fi, but has a wider range. It also avoids Wi-Fi-related security issues and proves to be more effective than cellular in most cases.

Monitoring systems should provide enormous amount of information/data about the performance of a solar system. Sensors can help monitor external weather conditions like radiance of the Sun, wind speed, temperature, and other electrical parameters such as string voltage, string current, power generated, number of units generated, battery voltages in case of off-grid systems, and so on. All these data sets can be monitored from the central location. Data received from the various components in the field devices will be sent to the IoT devices from where the data will be pushed into cloud servers. Through the data that is available in the cloud servers, various applications can be developed to access this data from anywhere, anytime, and from any device.

Solar panel efficiency that ranges between 15% and 20% was proven as the major obstacle to implement this renewable energy in a big way. Technologies such as solar and wind that mainly rely on the weather conditions are proven to be unreliable for maintain the grid stability. Therefore, intelligent IoT solutions with autonomous controls such as data acquisition, processing, control, and so on can be implemented to improve the overall efficiency of these types of energy resources. These types of autonomous controls will allow the utilisation of energy from these renewable energy resources to be used optimally along with the available other resources to improve the overall efficiency of the system. IoT devices will help in prediction of optimal direction, that is, tilt angle of solar panel in case of solar power plants or wind direction in case of wind forms to increase the efficiency of the plants by increasing the energy generation. Operators can adjust the tilt of solar panel or the direction of windmills accordingly to get extreme performance. Real-time data can be monitored and utilised to take appropriate actions using the data generated from IoT devices, which will avoid unnecessary losses in the system.

Examples for the applications of IoT devices in renewable energy resources automation include solar energy and wind energy to extract maximum out of the existing systems. In the case of windmills, the direction of wind is dynamic. Therefore, horizontal axis wind turbine (HAWT) direction needs to be changed at right angles to the wind to extract the maximum kinetic energy from the wind. This can be achieved by collecting the weather data from various IoT devices installed in the farm and sending the same to the wind turbine controller. Wind turbine orientation can be modified as per the wind direction to get maximum efficiency. IoT also has a relevant application in solar power plants. As the Sun's direction is not static and is continuously changing from dawn to dusk, solar panel direction needs to follow the movement of the Sun to absorb maximum energy from sunlight. To extract maximum power, solar panels need to be placed at right angles to the sunlight by monitoring the solar irradiation intensity data from sensors. This process can be automated using elevation and azimuth actuators to maximise the efficiency. This will improve the performance of the plant and reduce the operational burdens.

Smart grid has a lot of scope for the IoT applications. Especially from the utility side, they can install IoT devices at consumer premises and continually monitor the load patterns on time-to-time or real-time basis. This allows the utility to plan the resources optimally that avoid the ignition of costly fossil fuel resources. Customers can also allow the utilities to shut down the major loads at their premises using IoT applications as and when the grid gets overloaded against which they will get additional perks from the utility. This ultimately results in reduction in power wastage. Consumers can also significantly reduce their monthly electricity bills.

Even though there are many benefits of using IoT for renewable energy production, there are also a lot of hurdles when it comes to the implementation in the field. Initial investment of money is a major challenge. Although the cost of the devices was reduced if compared to previous years, renewable energy equipment are still on the higher side. Updating the existing grid with IoT devices that collect data and monitoring equipment can contribute to major investment costs associated with renewable energy-based processes. Depending on current trends, IoT-enabled energy production equipment may be adopted on a large scale if prices fall significantly in the future.

In Indian scenario, a community is identified and to be electrified, if 10% of its resident's houses get electricity. Though the number of community houses that have been electrified has gone up to 84%, some 410 lakh houses are without electrical power (Ashok Jhunjhunwala, Prabhjot Kaur, 2018). This clearly emphasises the need for solar installations where the people are deprived of accessing electricity. The system proposed by Quang Ha & Manh Duong Phung (2019) describes the application of IoT as its master to harmonise the operation of various auxiliary systems in a coordinated manner. The system utilises readings received from local sensors through IoT networks and online resources to make itself more fault-tolerant. The Sun's direction is not static; hence, incidence of sunrays continuously varies with time, and this will impact the efficiency of the solar power plant. As proposed by Vijaya Rama Raju V. & Divya Mereddy (2015), solar panels must face the Sun at right angles at every point of time, maximum efficiency can be achieved. IoT devices play a major role in achieving the same. Identification of the Sun's location is very important when it comes to tracking applications to extract maximum energy. There are five procedures proposed to identify for the Sun's location, over a period of 100 years starting from 2010, by Grena R. (2012). As per Singh G.K. (2013), the researchers are exploring the ways to utilise the solar energy, which is affordable, green, and in-exhaustive to its maximum extent for the present and future generations.

To effectively integrate smart home products into IoT-based cloud solutions, Stojkoska B.L.R. & Trivodaliev K.V. (2017) proposed an end-to-end framework that integrates various elements of IoT architectures/platforms. Identified procedures followed for analysing the data acquired is a major challenge, along with other challenges like protocols and devices with interoperability which are highly important for the implementation of IoT. The article mentioned by Vijaya Rama Raju., Ganapathi Raju, Shailaja & Sugandha (2020) provides an outline of various models and applications for estimating solar power and time-series, depending on the straightforward data acquired from solar plants. This is one of the most important applications for the data collected from various IoT devices deployed in the field. Solar power generation requires efficient monitoring and management for making net zero energy in green buildings. Manh, Michel & Quang (2017) introduced a reliable IoT-based control system for controlling and managing the flow of energy from solar plant in micro grids. For better control, the data include not only local measuring sensors but also real-time meteorological data from IoT sources. IoT-based Dependent Control System (DepCS) for continuous operation was proposed by Tran & Ha (2015) where regulator control loop was formed using an actuator and transmitter system. Processors used inside such a system can use feedback control procedures based on their own computational capabilities. For effective application of DepCS system communication channels between actuators and transmitters through IoT is vital. Future communication is completely with IoT, which can accommodate uniquely identifiable and dynamic users, computing facilities and devices that interact with unparalleled convenience and financial benefits.

In case of IoT applications, Internet Protocol-based communications are widely used to interact with each other. These types of communication channels are designed to meet the limitations of the sensor platform used by IoT applications, creating a communication base that can provide the energy needed for energy, reliability, and Internet connectivity (Granjal, Monteiro & Silva, 2015). There is a growing need to use signal-processing and pattern recognition techniques to identify and address faults in PV arrays automatically. Different topologies were proposed by Buddha et al. (2012) to identify a possible shading pattern and developed algorithms to suggest optimal configuration. Another important area that can utilise the recorded data was forecasting of generation, and this was dealt by Picault, Raison, Bacha, Casa & Aguilera (2010) by considering the field measurements to identify the module parameters. Various algorithms used for forecasting the photovoltaic energy generation in diverse weather conditions was discussed by Daliento et al. (2017). Suitable solar power plant monitoring systems were proposed by Han, Lee & Sang-Ha (2015), which utilises power line communication that send data to data loggers. Researchers such as Eduardo, Ricardo, Pedro, Sabino & Damian (2006) also proposed to use module-level DC–DC converter with MPPT that has communication capabilities using PLC with the central server.

In the present study, a low-cost IoT device that can collect the data from inverters and push it directly to cloud servers was proposed, which will avoid the data loss and improve the overall efficiency of the system.

5.3 APPLICATION OF IoT TO MONITOR THE SOLAR ENERGY RESOURCES

Solar energy resources in a bulk capacity are normally located in remote places. To extract the maximum power from the solar power plants and reliable functioning, real-time monitoring is essential. There are two ways in which monitoring can be done: online (remote) and manual. Manual monitoring involves the human intervention, where an assigned personal will record the readings on a regular interval of time by visiting the site. The major issue in collecting these readings is that they are not time synchronised when the size of the site is large. Though the process is simple, it will consume a lot of man hours, and the amount of information that can be read from the display units of the inverters is also limited. Normally, inverters display voltages, currents, power, and energy. Operator will not fetch the intrinsic details such as irradiation, temperature, histograms, and so on. As the power plant sites are located at remote locations or rooftops, manual method is cumbersome, expensive, and prone to human errors.

To enhance the performance of the plant and predict the generation, along with the monitoring of electrical parameters, one has to monitor weather parameters like irradiation and temperature. To overcome the disadvantages of manual monitoring, this process can be automated using IoT devices where the IoT devices are configured with the field sensors and collect the data without human intervention. The historical data sets gathered by these devices can be used to predict the generation from renewable energy resources and schedule the other power plants, accordingly, thus making the entire process of electrical energy generation more economical and environmental-friendly. Important infrastructure that is required for remote monitoring is communication channel. Different communication channels like WAN/LAN can be used for long distances. RS485/RS432 can be used in case of local monitoring. To push the data to cloud, IoT devices can be configured with the local Wi-Fi. An IoT module was developed in-house as per the requirements to gather various parameters and push that data to the cloud server. The performance of the plant can be monitored in a local/remote device by pulling the data from cloud.

5.3.1 MAJOR ADVANTAGES OF IoT IN RENEWABLE ENERGY APPLICATIONS

- **System Automation:** Major components of solar system include solar panels and inverters. In terms of numbers, solar panels are huge in number for commercial plants ranging from MWs to GWs. Monitoring of those panels poses a challenge to the operators. Monitoring and control need to be done at panel level and system level. Manual inspection becomes a cumbersome process. Installation of sensors and integrating them with IoT infrastructure will help them in reducing the costs and improve the system efficiency.

- **Economical Aspects of IoT:** IoT allows the system operators to continuously monitor the generation patterns of different sites located in geographically wide area from the remote location by the deployment of sensors at the solar installations.
- **Grid Stability:** One of the major obstacles in integration of solar energy sources into the grid is unpredictable nature of the availability of solar irradiation, and tamper stability of the grid. This challenge can be overcome by collecting the weather data using sensors that are integrated with IoT and predict the availability of irradiation. This data can be utilised by the system operators for automatic scheduling of other generating stations that utilise fossil fuels and maintain the stability of the grid.

5.3.2   IoT DEVICE FOR DATA ACQUISITION – A CASE STUDY

Indigenously developed IoT device was used to collect the data from the field sensors. The device consists of various components like RS 485 converter to communicate with the inverter using the MODBUS protocol in RTU mode to read various data points stored in the inverter such as voltage, current, power, and energy. The converter will send the data to microcontroller. Arduino Nano was used to acquire the data from the plant. Then, it will process it and send it to cloud using ESP 8266 module using the local Wi-Fi. The data have been collected in real time from the solar plant using IoT devices. IoT device used for data acquisition from solar power plant is shown in Figure 5.4. Components used to build the board include Arduino Nano at the core used for processing, to push the data received from the sensors to cloud ESP8266 Wi-Fi Module was used. MAX 485 is used for RS 485 communication between the IoT device and the inverter. 5V, 2A DC power supply was used to provide the necessary power for the board.

5.3.3   PROCEDURE TO ACQUIRE DATA FROM THE PLANT USING IoT SENSOR BOARD

Step 1: IoT sensor board installed at the distribution panel in the plant location will collect the data from the string inverters by using the MODBUS protocol. The device raises a query with the inverter for different electrical parameters such as currents, voltages, power, and energy.

Step 2: Inverter process the query and sends the requested information back to the device through the RS485 module.

Step 3: After the device receives the data from the inverter, it will push the data to the cloud server using the ESP8266 module. The ESP8266 module will be configured to the local Wi-Fi network.

Step 4: Once the data was uploaded to the cloud server, that information can be accessed from anywhere, anytime and any device. Different applications were

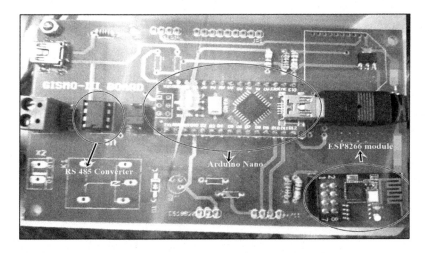

Figure 5.4 IoT device for data acquisition from solar power plant.

developed to monitor the data using digital gadgets such as mobile, laptop, and so on.

Step 5: Data stored in the cloud over a period can be used to develop the applications modules like energy prediction using artificial intelligence and machine learning algorithms (Figure 5.5).

5.3.4 SYSTEM ARCHITECTURE

System under study was a 100kWp solar power plant with the layout as shown in Figure 5.6. Entire plant was divided into five strings with each sting consisting of 80 modules. These strings are connected with the five numbers of 20kVA inverters each for voltage control and voltage conversion from DC to AC. Then, the five inverters are connected at the local LT distribution panel, where the IoT device was also installed. Wired communication channels were provided from the device to the individual inverters. Seasonal tilt provision is given to extract the maximum power.

5.4 ANALYSIS OF RESULTS

Any electrical infrastructure that was integrated with digital environment will offer maximum transparency of the system that allows the operator to monitor and act upon the events immediately. This helps the operator to leverage increased productivity and efficiency of the system. In the present study, IoT devices were deployed in the 100kWp rooftop system installed at 17.5203°N latitude and 78.3674°E longitude, that is, located on the Deccan Plateau in the northern part of South India. Generation of the plant was monitored over the years using the data collected from the IoT devices. Data received from the devices were pushed and stored in the cloud server. Monitoring of the data can be done from the central server by the system operator

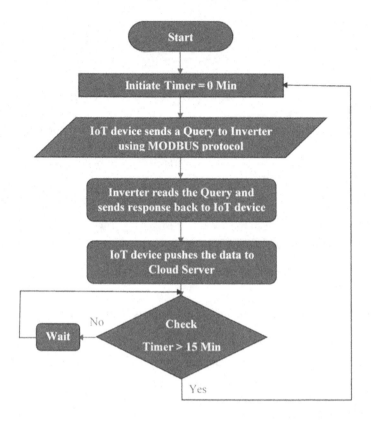

Figure 5.5   Data acquisition flow chart.

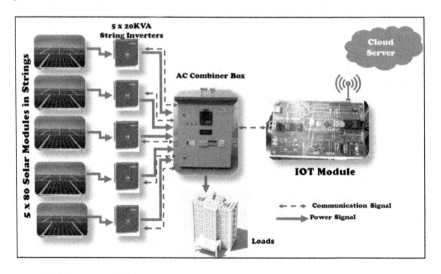

Figure 5.6   Architecture diagram.

from the central location. To give the freedom from location, a mobile application was also developed to monitor the data from anywhere and anytime and any device. This has created an ecosystem to address the issues arise in the plant within no time.

Data received from the plant were displayed in the dashboard in different formats as per the requirements by the operator. Operator can view the day in the real time by selecting the day wise display of the generation data as shown in Figure 5.3. As per the recorded data, it can be concluded that the generation starts normally at 06:00 hrs in the morning and continues till 18:00 hrs in the evening. Power generation follows the bell curve with peak at around 12:00 hrs. The dips in between the daily generation curve represents the fall in generation due to passing clouds on the plant. In case of any interruption, the generation comes to zero. This could be due to the grid failure. In case of grid failure generation will get restored immediately once the Diesel Generator (DG) starts. If there is interruption for more than 3–5 min, operator will get an alarm/message and will be alerted about the failure. At that time, the operator can take immediate action and address the issue, which was not possible without IoT applications. This way number of long interruptions can be converted into short interruptions. The typical generation profile shown in Figure 5.7 is also showing the number of units generated during the day by integrating the power generated during the day as 387 kWh. This clearly indicates that the number of units generated per kW is 4.8375 units, which is higher than the typical value of four units per kW per day which is also one of the performance indicators for the power plant. System operator can monitor at the end of the everyday and analyse the reasons for reduction in the performance of the power plant. There could be several reasons for reduction in generation like partial shadow due to nearby objects such as buildings, trees, or various civil structures, cloudy days during rainy season or

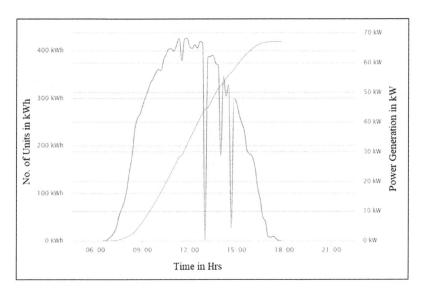

Figure 5.7 Solar power and energy monitored by IoT during a typical sunny day.

accumulated dust during summer season, and so on. Operator can analyse the reason for less generation and will be in a position to take necessary action on immediate basis.

Solar energy data given in Figure 5.8 represent the number of units generated on daily basis in a particular month. From the figure it is clearly evident that the generation is not consistent throughout the month; this mainly depends on the weather conditions and several other conditions. If there was a dip in the generation first, we need to check with the generation expected as per the irradiation on that day. If it does not match with the expected generation, then operator has to check for the faults in the system. Similarly, generation can be monitored over a period of year as shown in Figure 5.9. This clearly shows the seasonal variations in the generation throughout the year. Energy generation is peak during the early days of the spring season in India when we will have more sunny days at moderate temperature. The generation is lowest during the rainy season when we experience a greater number of cloudy and rainy days that diminishes the generation.

The system considered here is seasonal tilt system where the orientation of the panels was changed every 3 months throughout the year. Generation data over the years from the time of its installation are shown in Figure 5.10. This clearly shows that the generation is maximum in the year 2016 and the number of units generated were 0.161852 billion units, whereas the number of units generated during the year 2018 were only 0.122325 billion units. This reduction in the generation is due to the failure of one of the 20 kVA inverter. The data received from the IoT devices will clearly indicate the health of the system.

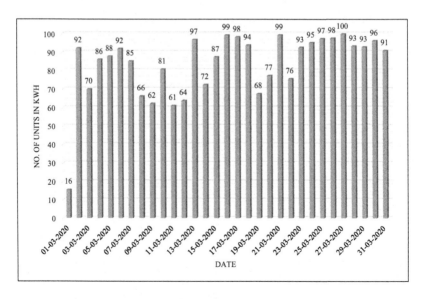

Figure 5.8   Solar energy generated during a Month.

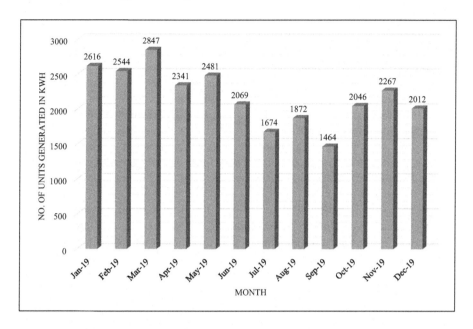

Figure 5.9 Solar power and energy generated during a year.

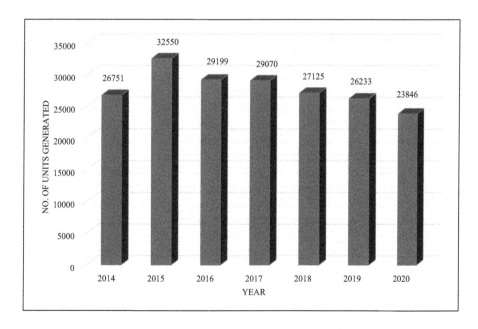

Figure 5.10 Database over a period of time: Year-wise generation of solar energy.

5.5 CONCLUSION

A completely new ecosystem of technology that effectively and efficiently improves day-to-day operations of human activities was created using IoT. This would extensively reduce the cost of living by automating the manual processes. For synchronous communication, it connects physical devices with Internet. Today, the cost of electricity is so high, and also dependency on fossil fuels challenges the mankind with global warming. Hence, there is a great need to use resources that can generate electricity without impacting the environment. This has emphasised to use solar energy to generate electricity using natural sunlight. Most of the solar power plants were installed across wide areas with numerous solar panels and multiple devices in each plant.

System operators must collect and process data from multiple devices in remote locations to achieve optimum power efficiency. IoT devices can eliminate the risks and challenges associated with traditional wire installations and make measuring and monitoring data easier and cheaper. This decentralised structure is a very simple, easy way to compare any expansion to a large solar composition. With an IoT-based solar panel monitoring system, system operators can view the performance metrics in real time, get optimised setup instructions, eliminate power outages in remote locations, reduce costs associated with 24/7 physical monitoring, ensuring optimal performance, analyse energy consumption and energy production while continuing to improve energy consumption and reduce waste, and monitor the working of various parts of solar devices and determine if they are in perfect condition.

5.6 FUTURE SCOPE

Solar energy is one of the most inexpensive, renewable, clean, and environmentally friendly energy sources in the world. As a result, many researchers have undergone an extensive study on solar energy to maximise the solar energy generation from solar radiation in a better and efficient manner. Though it has several advantages over non-renewable energy resources, as the generation majorly depends on the environmental conditions, it is not a completely dependable energy source by the system operators. Hence, many researchers have tried to develop models to predict the generation from the statistics. Recent advancements in the data science have paved ways to make use of different algorithms to solve prediction problems in a more efficient way. Various data mining methods can also be used to estimate solar energy. The application of data science algorithms requires huge amount of data, for training and testing of the models. Hence, the data received from the IoT devices will be definitely be useful in such applications. The outcomes of these research works will augment the system operators to plan the available resources in a much-optimised manner and to avoid firing costly secondary stations.

REFERENCES

Buddha, S., Braun, H., Krishnan, V., Tepedelenlioglu, C., Spanias, A., Yeider, T., Takehara, T. (2012). "Signal processing for photovoltaic applications," *IEEE ESPA*, pp. 115–118, 12–14 Jan. 2012.

Daliento, S., Chouder, A., Guerriero, P., Massi Pavan, A., Mellit, A., Moeini, R., Tricoli, P. (2017), "Monitoring, diagnosis, and power forecasting for photovoltaic fields: A review", *International Journal of Photoenergy* 2017, 1–13. Online publication date: 1–Jan–2017.

Granjal, J., Monteiro, E., Silva, J.S. (2015). "Security for the internet of things: A survey of existing protocols and open research issues", *IEEE Communications Surveys & Tutorials*, vol. 17, no. 3, pp. 1294–1312.

Grena, R. (2012). "Five new algorithms for the computation of sun position from 2010 to 2110", *Solar Energy*, vol. 86, no. 5, pp. 1323–1337.

Ha, Q., Phung, M.D. (2019). "IoT-enabled dependable control for solar energy harvesting in smart buildings", *IET Smart Cities*, vol. 1, no. 2, pp. 61–70.

Han, J., Lee, I., Sang-Ha, K. (2015). "User friendly monitoring system for residential PV system based on low-cost power line communication," *IEEE Transactions on Consumer Electronics*, vol. 61, no. 2, pp. 175–180.

Jhunjhunwala, A., Kaur, P. (2018). "Solar energy, dc distribution, and microgrids: Ensuring quality power in Rural India", *IEEE Electrification Magazine*, vol. 6, no. 4, pp. 32–39, Dec. 2018.

Phung, M.D., De La Villefromoy, M., Ha, Q. (2017). "Management of solar energy in microgrids using IoT-based dependable control", *2017 20th International Conference on Electrical Machines and Systems (ICEMS)*, Sydney, NSW, Australia. doi: 10.1109/ICEMS.2017.805644.

Picault, D., Raison, B., Bacha, S., de la Casa, J. and Aguilera, J.(2010), "Forecasting photovoltaic array power production subject to mismatch losses", *Solar Energy*, vol. 84, no. 7, pp. 1301–1309.

Roman, E., Alonso, R., Ibanez, P., Elorduizapatarietxe, S. and Goitia, D. (2006), "Intelligent PV module for grid-connected PV systems", *IEEE Transactions on Industrial Electronics*, vol. 53, no. 4, pp. 1066–1073, Aug. 2006.

Renewables (2019). Market analysis and forecast from 2019 to 2024, Fuel report — October 2019 by International Energy Agency.

Tran, T. and Ha, Q.P. (2015). "Dependable control systems with Internet of Things", *ISA Transactions*, vol. 59, pp. 303–313.

Singh, G.K. (2013). "Solar power generation by PV (photovoltaic) technology: A review", *Energy*, vol. 53, pp. 1–13.

Spanias, A.S. (2017). "Solar energy management as an Internet of Things (IoT) application", *8th International Conference on Information, Intelligence, Systems & Applications (IISA)*, Larnaca, Cyprus, 15 March 2018. doi: 10.1109/IISA.2017.8316460.

Stojkoska, B.L.R., Trivodaliev, K.V. (2017). "A review of internet of things for smart home: Challenges and solutions", *Journal of Cleaner Production*, vol. 140, pp. 1454–1464.

Vijaya Rama Raju, V., Mereddy, D. (2015). "Smart dual axes solar tracking system", *2015 IEEE International Conference on Energy Systems and Applications (ICESA 2015)* organized by Dr. D. Y. Patil Institute of Engineering and Technology, Pune, India 30 Oct – 01 Nov, 2015 sponsored by IEEE Pune section.

Vijaya Rama Raju, V., Ganapathi Raju, N. V., Shailaja, V., Padullaparti, S. (2020). "IoT based solar energy prophecy using RNN architecture", *E3S Web of Conferences*, Volume 184 (2020).

6 Machine Learning Techniques in IoT Applications: A State of The Art

Laxmi Shaw
Chaitanya Bharathi Institute of Technology
Hyderabad, India

Rudra Narayan Sahoo
TCS
Hyderabad, India

Hemachandran K.
Woxsen University
Hyderabad, India

Santosh Kumar Nanda
Techversant Infotech Pvt Ltd
Trivandrum, India

CONTENTS

6.1 Introduction 106
6.2 Materials and Methods 108
    6.2.1 Literature Review.. 108
6.3 Role of ML and DL in IoT applications 109
    6.3.1 ML for IoT.. 109
        6.3.1.1 Supervised Learning... 109
        6.3.1.2 Unsupervised Learning 110
        6.3.1.3 Semi-Supervised Learning 112
        6.3.1.4 Reinforcement Learning 112
    6.3.2 DL for IoT Application... 112
6.4 Issues and limitations in implementing ML and DL in IoT 113
6.5 Conclusion and Future Scope.. 114
References.. 115

6.1   INTRODUCTION

The present innovation of Internet of Things (IoT) provides intelligent and innovative services by being connected with numerous devices within network. The applications of IoTs have increased rapidly, it requires massive data storage, capable devices along with intelligent (Mahdavinejad et al., 2017) data processing techniques. The growth of the IoT can be recognized to convergence in advance took place over the earlier years in technology and communication. The influence of IoT shows the dominant control over humans, it has rapidly expanded. IoT devices assist us to pursue our regular activities which comprises smartphones, home assistants gadgets like Google-Play, automatic-driven vehicles, home computerization systems, smart elevators, room-temperature control systems, and unmanned aerial vehicles (UAVs) such as drones for ecological monitoring (Adi, Anwar, Baig, & Zeadally, 2020).

IoT gadgets and their associated platforms are utilized for a huge sending and receiving enormous volume of information through cloud computing. IoT gadgets are planned with routine rules that deem the asset constrained nature of these gadgets, to safeguard power utilization related with gadget operations. (Doshi, Apthorpe, & Feamster, 2018). The most well-known IoT application layer conventions are CoAP, MQTT, AMQP, and HTTP. (Yazici, Basurra, & Gaber, 2018). The MQTT and AMQP conventions are required persistent force flexibly or to reinforcement power source, when introduced to IoT gadgets. Those conventions are more force due to their empowered highlights of exchange of prolonged messages(Chin, Callaghan, & Lam, 2017). The main contribution of this chapter is to respond to the following demands:

- IoT's impacted in today's lifestyle
- Applications of IoT
- Machine learning models for IoT
- Uses of IoT devices in the current industry, organizations, and in different fields

IoT has got many applications; however, this covers smart household devices such as light-lamps, charging adapter, electric meter, fridge, microwave oven, air conditioner, temperature sensor, smoke-detecting sensor, Internet Protocol camera to more classy devices such as radio frequency-ID gadgets, heartbeat counters, accelerometer, parking-lot sensors, variety of other automobile sensors, and so on. The outdated data acquisition, storage and processing methods may not be appropriate for the comprehensive data engendered by the IoT devices. Therefore, for trends, behaviors, forecasts, and evaluation, the immense bulk of data can also be used. Furthermore, heterogeneity of IoT produced data is the reason for the existing mechanisms of data processing. Machine learning (ML) is ruminated in this sense to be one of the most suitable computing platforms to provide IoT devices with embedded knowledge.

ML is a subset of artificial intelligence (AI), which allows machines to learn from its experience without being explicitly programmed (Cui et al., 2018). For smart data analysis, frequently applied ML algorithms are mentioned in Table 6.1 and their applicability in IoT also discussed in the following Section 6.2.There are several IoT devices with many applications. To mention a few of them are as follows:

- Amazon Echo: This is an amazing device with a lot of smart features such as integrated voice reply, time alarms, live streaming, play music, making to-do lists, and so on. This device has personal voice-controlled assistant known as "Alexa" or "Echo".
- August doorbell cam: This IoT device is featured with vision and speech. When we are inside, Through phone, we can see and speak with our visitors from outside itself. It also has smart locks and 24 h video monitoring system.
- Awair: Awair is an amazing product used to monitor the CO_2 and humidity levels all the time. It is used to track the dust levels and VOCs in the air. It is mainly used for the people with allergies or hypersensitive reactions.
- Belkin WeMo: Belkin had launched a product in series named as WeMo to control home electronics remotely.
- Canary: It is a home security gadget which captures video, audio and sends alerts to your mobile or pc's. It identifies whoever standing before the entrance and that video will be streamed in your mobile.
- Elgato Eve: It is an Apple product developed to monitor inner house air moisture, outside weather condition, power consumption, and to check whether home windows and doors are opened or closed.

IoT is very efficient from multiple points of view. However, shockingly, this technology has not existed at this point, and it isn't completely secure. The entire IoT platform, from manufacturers to consumers, actually has abundant security hitches of IoT to persist, for example, manufacturing standards, update management, physical hardening, and users' knowledge and awareness. To overcome these IoT security issues, the following points need to be considered:

- Guard and safe the network connecting IoT devices by implementing antivirus, anti-malware and firewalls and authenticating the IoT devices using two-factor verification, biometrics and digital credentials.
- The connection between the IoT device and App will be secured using Data Encryption & Decryption.
- Usage of IoT Security Analytics to recognize IoT-specific intrusions and threats and usage of IoT API Authentication methods to guarantee that APIs are shared with only approved devices, developers and applications to recognize possible threats and attacks against individual APIs.
- To ensure their proper operation for their IoT applicatons, IoT system vendors need to conduct a large test of all the third-party components and modules they are using in their IoT products.
- The app developer must do a complete research on the security of their IoT applications and try their finest to incursion a perfect balance between the user interface and security of their IoT apps.
- Updating of the latest IoT threats and breaches regularly.

These solutions for IoT security issues can be taken into our mind when implementing machine learning with IoT. The ML techniques are the foundation of future prediction and are utilized in navigating to a hostile place where human expertise such as robotics, UAVs, Drones, and so on cannot be used. ML algorithms have come up with a dynamic solution where the problems are changing with time. In addition, they are utilized in practical smart systems and in Google by making use of ML to analyze threat in contradiction of android phone users and implementations running on android phones. Similarly using ML, Amazon has developed a facility called Macie to segregate data stored in its cloud storage. Deep learning (DL) is the subset of ML which is used to sort and classify millions of Facebook images and help to protect individual personalized accounts.

The rest of the chapter is structured as follows: Table 6.1 shows numerous case studies that have been used to address IoT provocation by utilizing ML models. We discuss the research gap and overview of the work done by IoT technology in Section 6.2. In Section 6.3, we discuss the motivation and objective of this survey followed by a brief review of state-of-the-art of ML and DL methods in the IoT devices. We review the active ML- and DL-based outputs for IoT devices in Section 6.4. Future research advices are discussed in Section 6.5, which are concluded at the end of this chapter.

6.2 MATERIALS AND METHODS

According to IoT Analytics estimates, there were roughly 9.5 billion connected IoT devices at the end of 2019, and numerous overviews have been distributed that spread to the various IoT networks. In this segment, we sum up the current studies and contrast them with our implementation. Supposedly, a large portion of the overviews we found in the literature doesn't concentrate on the ML methods utilized in IoT. Besides, the current studies are either implementation-explicit or doesn't incorporate the complete range of network and protection in the IoT systems. Table 6.1 sums up the current reviews in the survey that spreads to various parts of the IoT. In Table 6.1, we layout the subjects canvassed in these studies, and the individual improvements are incorporated in our review. Table 6.1 records a short synopsis of the evaluated works.

The current review covers the utilization of ML in IoT by considering the current customary arrangements and the goals which pay the way to new developing areas. Nonetheless, overviews covering ML and DL build arrangements which doesn't exist. Despite the fact that ML and DL have been appealed in a couple of studies, the general data on the systematic utilization of ML and DL is limited. To address the drawbacks, we conducted a complete study of ML and DL models utilized in IoT gadgets.

6.2.1 LITERATURE REVIEW

The ability of handheld devices to run ML models has been examined in the early plethora of the new century (Kargupta et al., 2004, 2002). The work has given proof of the capability of edge investigation, yet before the time of mobiles, TABs and PCs.

The dissemination of ML techniques implemented on android mobiles, with the ascent of IoT, has been concentrated completely in the territory of pocket data analysis. It is presently apparent that edge gadgets are reciprocal to distribute-computing to scale out ML frameworks. Moreover, the arrangement of DL techniques is a rising exploring region with anxiety of DL models through pruning and quantization (Anwar, Hwang, & Sung, 2017). The other methodology is to legitimately prepare tiny systems to diminish inactivity. Mobile-Nets models from Google is the methodology used. (Hussain, Anpalagan, Khwaja, & Naeem, 2017). Advances in IoT will direct a broad organization of DL models at the edge to decrease inertness. The DL techniques and conventional shallow learning techniques are feasible choices for edge gadgets, particularly those with quality necessities. Table 6.1 describes the frequently used ML models for smart data analysis in IoT, and Table 6.2 gives a clear review of how ML plays a vital role in IoT.

6.3   ROLE OF ML AND DL IN IoT APPLICATIONS

The implementation of ML models on IoT devices affords good solutions for numerous problems that the IoT devices can conquer with the correct outfit, and it can also tackle current and future real-world challenges.

6.3.1   ML FOR IoT

We're in reality all acquainted with ML in our regular day-to-day existences. Both Amazon and Netflix use ML to get familiar with our inclinations and give a superior encounter to the client. That could mean recommending items that you may like or giving pertinent suggestions for films and TV shows.

In addition, in IoT, ML can be incredibly important in forming our current circumstances to our own inclinations. The Nest Thermostat is an incredible model. It utilizes ML to become familiar with your inclinations for warming and cooling, ensuring that the house is the correct temperature when you return home from work or when you get up toward the beginning of the day.

The ML algorithms are categorized into four learning models: supervised learning, unsupervised learning, semi-supervised learning, and reinforcement learning (Yazici et al., 2018). The types of ML algorithms are given in detail in the following section. Figure 6.1 represents the features of the IoT. The details are briefly mentioned in the following sections.

6.3.1.1   Supervised Learning

In this supervised ML algorithm, output variable (Y) is dependent on input variables (X).

$$Y = f(x) \qquad (6.1)$$

Different supervised learning algorithms are utilized to apply some functions on 'X' to build models to find 'Y'. The supervised ML algorithm can be used for two kinds

Table 6.1

Algorithms and Their Applications

References	Machine Learning Algorithms	Data Processing Objective
(Gao & Thamilarasu, 2017), (Outchakoucht, ES-SAMAALI, & Philippe, 2017)	Naïve Bayes	Classification
Doshi et al. (2018)	K- Nearest Neighbours (KNN)	Classification
(Hussain et al., 2017)	K- Means Clustering	Prediction
(Ham, Kim, Kim, & Choi, 2014)	Support Vector Machine (SVM)	Classification
(Xuanxuan, 2018)	Linear regression	Prediction /Regression
(An et al., 2017)	Principal component Analysis (PCA)	Classification and Clustering
(Hussain et al., 2017)	Q-Learning	Reinforcement Learning
(Azmoodeh, Dehghantanha, & Choo, 2019)	Deep Learning(DL)	Classification
(Chauhan et al., 2018), (HaddadPajouh, Dehghantanha, Khayami, & Choo, 2018), (Karbab, Debbabi, Derhab, & Mouheb, 2018),(Su et al., 2018)	Recurrent Neural Network (RNN)	Classification and Prediction
(Lakshmanaprabu et al., 2019)	Random Forest and Decision Tree (DT)	Classification /Regression
(Kaminski et al., 2017)	Feed Forward Neural Network (FFNN)	Classification and prediction/Clustering
(Petladwala, Ishii, Sendoda, & Kondo, 2019),(Lavrova, Pechenkin, & Gluhov, 2015)	Canonical Correlation Analysis (CCA)	Detection and Prediction
(Zhang, Li, & Wang, 2019)	Deep Belief Network (DBN) and Genetic Algorithms(GA)	Prediction
(Min et al., 2019)	Reinforcement Learning(RL)	Prediction
(Zou, Jiang, Lu, & Xie, 2014)	Extreme Learning Ma-chine(ELM)	Prediction

of problems, and they are classified for target label and regression for quantitative prediction.

6.3.1.2 Unsupervised Learning

In this algorithm, the output variable (Y) is not dependent on input variables (X).

$$Y = f(x) \tag{6.2}$$

Here, various unsupervised learning algorithms are used to construct a group of labels based on some similarities to find 'Y', and the unsupervised learning algorithms are sub-divided into Clustering and Association, applying some functions on 'X'.

Table 6.2

Existing Literature of ML in IoT

Machine Learning Algorithm	Objective	Smart Features Use Cases	References
Principal Component Analysis(PCA)	Fault detection	Monitoring Public Places	(Monekosso & Remagnino, 2012)
Canonical Correlation Analysis (CCA)	Fault detection	Monitoring Public Places	(Monekosso & Remagnino, 2012)
Support Vector Machine (SVM)	Classify data Real time application	Almost all use cases	(Khan, Khan, Khan, & Anwar, 2014)
Feed Forward Neural Network (FFNN)	Reducing energy consumption, Forecast the states of elements	Health monitoring	(Ramalho, Neto, Santos, Filho, & Agoulmine, 2015), (Kotenko, Saenko, Skorik, & Bushuev, 2015)
K- Means	Outlier Detection, fraud detection, Analyzing small data set, forecasting energy consumption, Passengers travel Pattern	Smart City, Home, Citizen, Controlling Air and Traffic	(Shukla, Kosta, & Chauhan, 2015), (Tao & Ji, 2014),(Hromic et al., 2015), (Souza & Amazonas, 2015), (Ma, Wu, Wang, Chen, & Liu, 2013)
Naïve Bayes	Food Safety, Passengers Travel Pattern, Estimate the numbers of nodes	Smart Agriculture, Citizen	(Ma et al., 2013)
K- Nearest Neighbour (KNN)	Passengers' travel pattern,	Smart citizen	(Ma et al., 2013),(Do, Douzal, Marie, & Rombaut, 2015)
Density-based clustering	Fraud detection, Analyze small data set	Smart citizen	(Shukla et al., 2015), (Ma et al., 2013), (Khan et al., 2014)
Classification Regression trees Clustering	Traffic Prediction, Increase data abbreviation, Weather prediction	Smart Traffic, Smart citizens, Anomaly detection, Weather forecasting, market analysis	(Qin et al., 2016), (Yogita & Toshniwal, 2013),(Jakkula & Cook, 2010),(Ni, Zhang, & Ji, 2014)

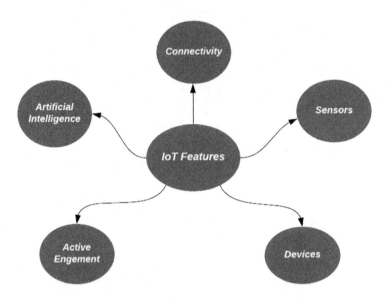

Figure 6.1   Features of IoT.

6.3.1.3   Semi-Supervised Learning

The input variables (X) are usable in semi-supervised learning algorithms with a mixture of labeled and unlabeled methods. The mixture of methods of guided and unsupervised learning is used to find 'Y'.

6.3.1.4   Reinforcement Learning

Machines/software agents may make optimum choices for a given situation to optimize the output of a system based on forward/feedback.

6.3.2   DL FOR IoT APPLICATION

DL is all over the place. This subset of Artificial Intelligence curates your online media and serves your Google indexed lists. Long before, DL could likewise check your vitals or set your thermostat. MIT scientists have built up a framework that could bring DL neural networks to new and a lot more modest places, similar to the smaller microchips in wearable clinical gadgets, household appliances, and the 250 billion different items that comprise the "Internet of things" (IoT).

The framework, called MCUNet, plans smaller neural networks that convey remarkable speed and exactness for DL on IoT gadgets, in spite of restricted memory and processing power. The innovation could encourage the extension of the IoT universe while sparing energy and improving information security.

DL is a subset of ML skill which comes from the artificial neural networks (ANN). The DL is useful for high volume data that is going to be used for training purposes to achieve better results. The profound deep neural network (DNN) system consists of neurons which are measured as variables and associated through weighted connections measured as parameters. To reach the goal, learning methodologies are aligned with the system. The training is done by utilizing labelled and unlabelled data from different learning methods, separately trailed by the step change of the weights between every set of neurons. In this manner, when we examine Deep Learning, we allude to enormous DNN systems where the word deep alludes to the quantity of surfaces in that network (Mohammadi, Al-Fuqaha, Sorour, & Guizani, 2018),(Yao et al., 2018). Figure 6.2 represents the pictorial representation of the current status of IoT and its future perspective.

6.4 ISSUES AND LIMITATIONS IN IMPLEMENTING ML AND DL IN IoT

Since the DL methodologies are very application-oriented, one prepared model may not be appropriate for other comparative issues. It requires a huge amount of respective data to retrain the model (L'Heureux, Grolinger, Elyamany, & Capretz, 2017). This traditional method will not be an issue for various static setups. However, the current uses of IoT such setups will not be candid for utilize.

As all the neural network models will act like a black-box mimic the humanoid brain nature, it is difficult to observe about humanoid brain senses and work accordingly. Similarly, how the DNN setup will reach the goal by processing the input parameters is also not clear sometimes. Therefore, it is tough to predict the next state

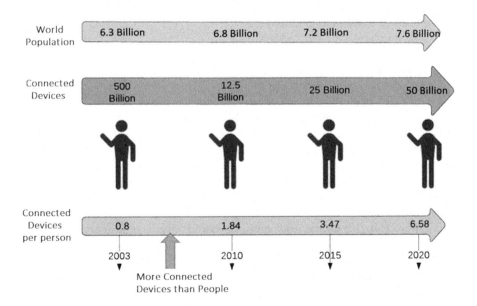

Figure 6.2 Current status and future perspective of IoT.

to reach the goal, and therefore it can't be utilized for those applications. Furthermore, the IoT devices and the data generated from the sensors are very massive, diversified, inconsistent in speed, and uncertain. Thus, the majority of the conventional ML models are not proficient and scalable as much as necessary to deal with Internet of Things data. Therefore, significant changes are needed. Apart from that, ML models have some unpredictable issues, for example, memory size, computational model, and test multifaceted nature. Likewise, conventional ML approaches need adaptability and are just restricted to low-dimensional problems. IoT gadget is tiny and regularly has energy imperatives with restricted processing power. Hence, direct use of traditional ML methods isn't reasonable in asset obliged situations. Then again, elegant IoT gadgets require real-time data analysis for ongoing applications, while conventional ML models are not intended to deal with consistent floods of information in real time.

ML expects factual properties over the whole data set continue as before, needs data pre-handling and removing of unwanted data from the data set before fixing it in to a particular model. However, that isn't the situation in true where data set from different sources have distinctive representations and formats. Besides, there may be contrasts among various pieces of the equivalent data set. This circumstance causes troubles for the ML model because the designed models are typically not intended to deal with semantically and grammatically different data. This advocates for effective answers for the heterogeneity issue. As the IoT devices can take the welfare of edge computing, and furthermore, ML and DL for IoT uses will expand the applications and environmental space. In any case, it will be difficult to execute DL methods in the edge gadgets. Moreover, the required time for training a DNN system assumes a significant part where time-based and current trending implementations. It would not be ready to exploit DL in the edge.

6.5 CONCLUSION AND FUTURE SCOPE

ML is the key innovation in IoT. These two are the evergreen technologies. ML patterns help to convey investigation for IoT concurrent applications. The ongoing rush for achieving ML for administrating system, there is an acute shortage of ML literature reviews about its concurrent implementations in IoT enabled-assistances and frameworks. Since the network plays a crucial function in the entire net, where ML and industrial IoT plays a vital role. This chapter predominantly accentuates the utilization of ML in IoT and the inclusion of ongoing advances. Because of the flexibility and advancing nature of IoT, it is difficult to cover every single concurrent application. This chapter has made an endeavour to cover significant utilizations of ML for IoT and the important technical procedures, including traffic violation picture capture. IoT gadget ID's, protection (security purpose), computer vision, networking dependent and other various concurrent applications of IoT. We have introduced an intensive report on the ongoing research project which utilizes ML for IoT, its advancement, and implementation in different areas. We have additionally introduced brief innovative research approaches that are basic to the ML of IoT.

REFERENCES

Adi, E., Anwar, A., Baig, Z., & Zeadally, S. (2020, 10). Machine learning and data analytics for the IoT. *Neural Computing and Applications, 32.*

An, N., Duff, A., Naik, G., Faloutsos, M., Weber, S., & Mancoridis, S. (2017). Behavioral anomaly detection of malware on home routers. In *2017 12th International Conference on Malicious and Unwanted Software (MALWARE)* (p. 47–54).

Anwar, S., Hwang, K., & Sung, W. (2017, 2). Structured Pruning of Deep Convolutional Neural Networks. *ACM Journal on Emerging Technologies in Computing Systems, 13.*

Azmoodeh, A., Dehghantanha, A., & Choo, K. R. (2019). Robust Malware Detection for Internet of (Battlefield) Things Devices Using Deep Eigenspace Learning. *IEEE Transactions on Sustainable Computing, 4* (1), 88–95.

Chauhan, J., Seneviratne, S., Hu, Y., Misra, A., Seneviratne, A., & Lee, Y. (2018). Breathing- Based Authentication on Resource-Constrained IoT Devices using Recurrent Neural Net- works. *Computer, 51* (5), 60–67.

Chin, J., Callaghan, V., & Lam, I. (2017). Understanding and personalising smart city services using machine learning, The Internet-of-Things and Big Data. In *2017 IEEE 26th International Symposium on Industrial Electronics (ISIE)* (p. 2050–2055).

Cui, L., Yang, S., Chen, F., Ming, Z., Lu, N., & Qin, J. (2018, 08). A survey on application of machine learning for Internet of Things. *International Journal of Machine Learning and Cybernetics, 9.*

Do, C. T., Douzal, A., Marie, S., & Rombaut, M. (2015, 09). Multiple Metric Learning for large margin kNN Classification of time series. In *2015 23rd European Signal Processing Conference (EUSIPCO).*

Doshi, R., Apthorpe, N., & Feamster, N. (2018). Machine Learning DDoS Detection for Consumer Internet of Things Devices. In *2018 IEEE Security and Privacy Workshops (SPW)* (p. 29–35).

Gao, S., & Thamilarasu, G. (2017). Machine-Learning Classifiers for Security in Connected Medical Devices. In *2017 26th International Conference on Computer Communication and Networks (ICCCN)* (p. 1–5).

HaddadPajouh, H., Dehghantanha, A., Khayami, R., & Choo, K.-K. R. (2018). A deep Recurrent Neural Network based approach for Internet of Things malware threat hunting. *Future Generation Computer Systems, 85,* 88–96. Retrieved from http://www.sciencedirect.com/science/article/pii/S0167739X1732486X

Ham, H.-S., Kim, H.-H., Kim, M.-S., & Choi, M.-J. (2014, 09). Linear SVM-Based Android Malware Detection for Reliable IoT Services. *Journal of Applied Mathematics, 2014,* 1– 10.

Hromic, H., Le Phuoc, D., Serrano, M., Antonić, A., Žarko, I. P., Hayes, C., & Decker, S. (2015). Real time analysis of sensor data for the Internet of Things by means of clustering and event processing. In *2015 IEEE International Conference on Communications (ICC)* (p. 685–691).

Hussain, F., Anpalagan, A., Khwaja, A. S., & Naeem, M. (2017). Resource allocation and congestion control in clustered M2M communication using Q-learning. *Transactions on Emerging Telecommunications Technologies, 28* (4), e3039. Retrieved from https://onlinelibrary.wiley.com/doi/abs/10.1002/ett.3039

Jakkula, V., & Cook, D. (2010). Outlier Detection in Smart Environment Structured Power Datasets. In *2010 Sixth International Conference on Intelligent Environments* (p. 29–33).

Kaminski, N., Macaluso, I., Di Pascale, E., Nag, A., Brady, J., Kelly, M., . . . Doyle, L. (2017). A neural-network-based realization of in-network computation for the Internet of Things. In *2017 IEEE International Conference on Communications (ICC)* (p. 1–6).

Karbab, E. B., Debbabi, M., Derhab, A., & Mouheb, D. (2018). MalDozer: Automatic framework for android malware detection using deep learning. *Digital Investigation, 24,* S48–S59. Retrieved from http://www.sciencedirect.com/science/article/pii/S1742287618300392

Kargupta, H., Bhargava, R., Liu, K., Powers, M., Blair, P., Bushra, S., . . . Handy, D. (2004, 04). VEDAS: A Mobile and Distributed Data Stream Mining System for Real-Time Vehicle Monitoring.

Kargupta, H., Park, B.-h., Pittie, S., Liu, L., Kushraj, D., & Sarkar, K. (2002, 04). MobiMine: Monitoring the Stock Market from a PDA. *ACM SIGKDD Explorations, 3.*

Khan, M. A., Khan, A., Khan, M. N., & Anwar, S. (2014). A novel learning method to classify data streams in the internet of things. In *2014 National Software Engineering Conference* (p. 61–66).

Kotenko, I., Saenko, I., Skorik, F., & Bushuev, S. (2015). Neural network approach to forecast the state of the Internet of Things elements. In *2015 XVIII International Conference on Soft Computing and Measurements (SCM)* (p. 133–135).

Lakshmanaprabu, S., Shankar, D., .M, I., Nasir, A. W., Varadarajan, V., & Chilamkurti, N. (2019, 10). Random forest for big data classification in the internet of things using optimal features. *International Journal of Machine Learning and Cybernetics, 10.*

Lavrova, D., Pechenkin, A., & Gluhov, V. (2015, 12). Applying correlation analysis methods to control flow violation detection in the internet of things. *Automatic Control and Computer Sciences, 49,* 735–740.

L'Heureux, A., Grolinger, K., Elyamany, H. F., & Capretz, M. A. M. (2017). Machine Learning With Big Data: Challenges and Approaches. *IEEE Access, 5,* 7776–7797.

Ma, X., Wu, Y.-J., Wang, Y., Chen, F., & Liu, J. (2013, 11). Mining smart card data for transit riders' travel patterns. *Transportation Research Part C: Emerging Technologies, 36,* 1—12.

Mahdavinejad, M., Rezvan, M., Barekatain, M., Adibi, P., Barnaghi, P., & Sheth, A. (2017, 10). Machine learning for Internet of Things data analysis: A survey. *Digital Communications and Networks.*

Min, M., Xiao, L., Chen, Y., Cheng, P., Wu, D., & Zhuang, W. (2019). Learning-Based Computation Offloading for IoT Devices With Energy Harvesting. *IEEE Transactions on Vehicular Technology, 68* (2), 1930–1941.

Mohammadi, M., Al-Fuqaha, A., Sorour, S., & Guizani, M. (2018). Deep Learning for IoT Big Data and Streaming Analytics: A Survey. *IEEE Communications Surveys Tutorials, 20* (4), 2923–2960.

Monekosso, D., & Remagnino, P. (2012, 01). Data reconciliation in a smart home sensor network. *Expert Systems with Applications, 40.*

Ni, P., Zhang, C., & Ji, Y. (2014). A hybrid method for short-term sensor data forecasting in Internet of Things. In *2014 11th International Conference on Fuzzy Systems and Knowledge Discovery (FSKD)* (p. 369–373).

Outchakoucht, A., ES-SAMAALI, H., & Philippe, J. (2017, 01). Dynamic Access Control Policy based on Blockchain and Machine Learning for the Internet of Things. *International Journal of Advanced Computer Science and Applications, 8.*

Petladwala, M., Ishii, Y., Sendoda, M., & Kondo, R. (2019). Canonical Correlation Based Feature Extraction with Application to Anomaly Detection in Electric Appliances. In *ICASSP 2019–2019 IEEE International Conference on Acoustics, Speech and Signal Processing (ICASSP)* (p. 2737–2741).

Qin, Y., Sheng, Q., Falkner, N., Dustdar, S., Wang, H., & Vasilakos, A. (2016, 02). When Things Matter: A Survey on Data-Centric Internet of Things. *Journal of Network and Computer Applications, 64.*

Ramalho, F., Neto, A., Santos, K., Filho, J. B., & Agoulmine, N. (2015). Enhancing ehealth smart applications: A fog-enabled approach. In *2015 17th International Conference on E-health Networking, Application Services (HealthCom)* (p. 323-328).

Shukla, M., Kosta, Y. P., & Chauhan, P. (2015). Analysis and evaluation of outlier detection algorithms in data streams. In *2015 International Conference on Computer, Communication and Control (IC4)* (p. 1–8).

Souza, A., & Amazonas, J. (2015, 12). An Outlier Detect Algorithm using Big Data Processing and Internet of Things Architecture. *Procedia Computer Science, 52,* 1010–1015.

Su, J., Vasconcellos, D. V., Prasad, S., Sgandurra, D., Feng, Y., & Sakurai, K. (2018). Lightweight Classification of IoT Malware Based on Image Recognition. In *2018 IEEE 42nd Annual Computer Software and Applications Conference (COMPSAC)* (Vol. 02, p. 664–669).

Tao, X., & Ji, C. (2014). Clustering massive small data for IOT. In *The 2014 2nd International Conference on Systems and Informatics (ICSAI 2014)* (p. 974–978).

Xuanxuan, Z. (2018, 07). Multivariate Linear Regression Analysis on Online Image Study for IoT. *Cognitive Systems Research, 52.*

Yao, S., Zhao, Y., Zhang, A., Hu, S., Shao, H., Zhang, C., . . . Abdelzaher, T. (2018). Deep Learning for the Internet of Things. *Computer, 51* (5), 32–41.

Yazici, M., Basurra, S., & Gaber, M. (2018, 09). Edge Machine Learning: Enabling Smart Internet of Things Applications. *Big Data and Cognitive Computing, 2,* 26.

Yogita, & Toshniwal, D. (2013). Clustering techniques for streaming data-a survey. In *2013 3rd IEEE International Advance Computing Conference (IACC)* (p. 951–956).

Zhang, Y., Li, P., & Wang, X. (2019). Intrusion Detection for IoT Based on Improved Genetic Algorithm and Deep Belief Network. *IEEE Access, 7,* 31711–31722.

Zou, H., Jiang, H., Lu, X., & Xie, L. (2014). An online sequential extreme learning machine approach to wifi based indoor positioning. In *2014 IEEE World Forum on Internet of Things (WF-IoT)* (p. 111–116).

7 The Farmer's Support System: IoT in Agriculture

Jitendra Kumar Rout, Bhagyashree Mohanty,
Shruti Priya, Pankhuri Mehrotra, and
Nidhi Bhattacherjee
KIIT Deemed to be University
Bhubaneswar, India

CONTENTS

7.1 Introduction 119
7.2 Related Work 121
7.3 The Objective of the Invention 123
7.4 The Farmer's Support System 124
    7.4.1 Tactile Component.. 125
    7.4.2 Predictive Component... 126
        7.4.2.1 Dataset Used... 126
    7.4.3 Vista Component 127
    7.4.4 Test Cases Design 127
    7.4.5 System Testing... 129
7.5 Results and Discussion 129
    7.5.1 Output from the Tactile Component 129
    7.5.2 Output from the Predictive Component... 130
    7.5.3 Output from the Vista Component... 131
7.6 Conclusion and Future Scope 140
References... 140

7.1 INTRODUCTION

In India, about two-thirds of the population earns a living through agriculture, especially in rural areas. Also, moving to get a job in different sectors is difficult due to the lack of employment opportunities as the working population is increasing exponentially. According to the PRS Legislative Research Economic Survey (2019), the agricultural growth in India has been fluctuating incessantly. It increased from −0.2% in 2014–2015 to 6.3% in 2016–2017 but again declined to 2.8% in 2019–2020. The gross value added (GVA) of 17.7% in 2013–2014 decreased to 15.2% in

2017–2018. The pressure exerted on the economy of a country like India with such a fast-growing population is huge, which results in a rapid increase in the demand for food production. Therefore, agriculture has become one of the most significant aspects when it comes to the economy and the development of the country. Consequently, farmers have become an important factor in the development of the agriculture sector. However, nowadays, due to excessive crop failures, farmers are opting to quit their farming work and move to get other small paid jobs for survival. Some farmers with no other options left are committing suicide due to the burden of loans and failure of crop harvest.

To stop such horrendous situations from occurring where we won't even have enough food production for survival, we need to support the farmers to maximize their crop yields. With the aim to support the farmers in increasing the crop production from the initial stage itself, we propose *The Farmer's Support System*, a model incorporating various technologies such as Internet of Things (IoT), ML, Data Analytics and android to provide prior knowledge about the crop success to the farmers.

The proposed system uses various different sensors using IoT technology to collect data such as soil moisture, light intensity, rainfall, temperature, and so on and send this data to a database for further analysis. Data analytic is used to find a suitable data set for applying a Deep Learning (DL) model with suitable and available attributes and the relationship between them. The DL model which has been used is Convolutional Neural Network (CNN). **Specific input from farmer via android app, input from sensors as well as data from database collectively used for training of DL algorithm. Subsequently, the predicted output is sent to the android app at farmers end for display.** Android is used in this system to provide better understanding and ease of usage for the farmers. After hosting and configuring our model, the required input for prediction is provided by the user which is then sent to the database. The model will do all its predictions and shows the output indicating the success or the failure of the crop along with some other information like the possibilities of rainfall, temperature, and moisture level through android. Also, according to the type of soil, it will provide an intellect regarding the crops that are suitable for such soil which will give high yields. Figure 7.1 shows the flow (sequences) of all the components present in *The Farmer's Support System.*

The field of invention relates to providing farmers with a support system and application so that they can maximize the crop harvest yield using IoT, Data Analytics, ML, and Android. In particular, our system provides an application through which a farmer depending on the soil type can know whether or not the selected crop will be successful, and in addition to this, intellect about the crop suitable for the mentioned soil is also given to help the farmer with better decision-making.

The rest of the chapter is organized as follows: details of the related works are discussed in Section 7.2. Section 7.3 is about the objective of the research done. The details of the farmer's support system are given in Section 7.4. The test cases are depicted in Section 7.5. Discussion of results is given in Section 7.6. Finally, Section 7.7 concludes the work with a note on possible future directions.

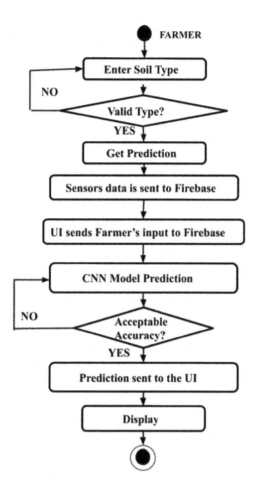

Figure 7.1   Activity diagram showing the flow in The Farmer's Support System.

7.2   RELATED WORK

Owing to the favorable agro-climatic zones and huge diversity in the land and soil types, India is ranked second for its agricultural outputs. However, during recent times, we have seen a lot of diversification and the severe consequences on the lives of farmers due to crop failures. This opened the ways to adapt to technologies in addition to the traditional farming methods to increase the yields. A number of inventions have been made to support farmers in agricultural activities. The following research articles and web references propose methods/devices to provide a helping mechanism to farmers in agricultural activities.

The work carried out by Dumont (2015) was particularly related to the efficiency of judging the balance amount of fertilizers that will lead to high yields as well. It calculates the amount of nitrogen phosphorus and potassium that should be absorbed based on the output of plant crops. **Though the invention is quite helpful as it can calculate the quantity of soil nutrients required in terms of N, P, K based on which the accurate fertilizer amount can be decided for better yield, but it may be very difficult for farmers to afford the amount of fertilizer anticipated.**

Aspelin (2001) devised a predictive device that monitors the growth and development of plants, prognosticate the need for plant protection through weed control steps and estimate the use of correct active agents. Incorporating a device to judge the removal of weed by simple farmers who have years of experience in controlling weed will become complicated for farmers during the agricultural process. Moreover, the additional active agents required may not be affordable for the poor farmers of the rural areas.

Schildroth (2018) developed an agronomic system which consists of two components: the first component includes information gathering and a network interface for receiving the agricultural prescription, and the second component is made with configuration to communicate and to output at least one agricultural action. Considering the complexity of the system and also the restriction of proper interfacing in rural areas, the system might fail to provide support when needed the most by farmers.

Pongnumkul (2015) discussed the significance of the use of cell phone sensors in agribusiness. These smart phones open a large number of opportunities for farmers because these smart phones are fully furnished with unlimited sensors. These farmers previously had limited access to all those advanced technologies and government new policies and schemes. In the interim, farmers on large-scale farms, who previously used assistance, would now be able to use cell phone sensors to upgrade profitability. Utilizing sensors implanted in cell phones can be inefficient due to water or falling down while farmers are working can damage the smart phone and replacing the smart phones, again and again, is not cost-effective for farmers.

Through research, Limbore (2015) manifested that to enhance investment and achieve a sustained increase in production, coherent and integrated long-term strategies and policies are required to cut back risk aversion and build flexibility among Indian rural producers. There's a necessity to produce remunerative prices for farmers to extend the incomes of farmers. The basic objective is to review the important agricultural crops production, export, and import of wheat crops. Even though this chapter is efficient enough to understand the importance of crop production, yet it does not mention any methodology to help farmers increase their crop yields besides it only focuses on a single crop (i.e. wheat).

Patil (2002) debated over utilizing the different occasions in the Indian scenario, that is, agribusiness (arrangements, procedures, structure), cultivating (real creation of yields in ranches), food, and nourishment (or unhealthiness). The choice of India is purposeful as shortly it may move to an "Evergreen Revolution". It's a hotbed of late ideas with minor upsets in horticulture/cultivating and represents a majority of economically developing countries in their progress toward food and nutritional security and decreasing destitution. This chapter describes the importance of involving technology in the agricultural process to increase productivity, but no specific mechanism is mentioned regarding how to use the technology to increase the yield or help farmers through it.

The objective of the findings by Kaur (2016) is to extend the cultivating range by posting and assessing various utilizations of ML in Indian horticulture and to enable the ranchers to propel their workup by numerous scores. The application of ML can surely be fruitful to the agricultural process but the diversity of land, soil type and climate make the attribute difficult to be chosen while prediction. Besides, the methodology for how this is going to reach farmers and will be understandable to them is not specified.

Rajeswari (2018) discourses the prediction of the most profitable crop that can be grown in agricultural land using ML techniques. It also uses an android system that will provide real-time crop analysis using various weather station reports and soil quality. Many areas with network restrictions set a drawback when the system tries to access data from the weather stations for prediction purposes.This uses random forest algorithms and is void of cost predictions.

Liakos (2018) provides a complete review of research dedicated to applications of ML in agricultural production systems. The review is about crop management (which includes applications for yield forecast, illness identification, weed discovery, crop quality, and species recognition), livestock management (which includes applications on animal welfare and livestock production), water management, and soil management. The key focus is to make profit in firming by applying ML. All the applications mentioned in the chapter address the vital factors in the agriculture process, but the ease of understanding and usability for farmers with minimal educational qualification makes it complicated.

Priya (2018) has used the well-known supervised ML algorithm, namely, Random Forest, to forecast crop yield depending on existing data to help farmers before cultivating the agricultural field. Though this model gives accurate results, it is limited only to Tamil Nadu which has a specific soil and weather conditions which are not same in other parts of a diverse and vast country like India.

The work by Devika (2018) is based on gathering agrarian information that can be stored and examined further for expectation and gauzing of harvest yield. Various data mining procedures like KNN and linear regression were used to foresee the harvest yield and to help farmers in picking the most appropriate yield. Though this method is easy to implement, yet the integration of the predictive model with the user interface becomes difficult and complex, thereby making the model tedious to use for farmers.

Apart from these few more works (Singh et al. (2010), Nagini et al. (2016), Shakoor et al. (2017), Ramesh and Vardhan (2015), Rajeswari et al. (2019), Nigam et al. (2019), Kumar et al. (2015), Salpekar (2019), Mishra et al. (2018))have been reported in the literature that has been done for crop yield prediction using different data mining and/or machine learning techniques.

7.3   THE OBJECTIVE OF THE INVENTION

As per the above findings and literature survey, there is still scope to provide support and helping methodologies to the farmer considering the shortcomings of the prior inventions related to the field of agriculture. Based on our key findings, we have proposed our model *The Farmer's Support System* in order to

- provide an easy-to-use application for farmers to gain essential knowledge prior to sowing the crop seeds.
- speculate whether the crop selected based on the soil type will be successful or not.
- give additional information about what crops should be sown on the particular soil type.
- provide additional intellect about weather and soil conditions for better decision making by the farmer.

7.4   THE FARMER'S SUPPORT SYSTEM

In adherence to our survey, none of the above web references, patents, and research articles directly relate to our invention, and we need an easy to use responsive system that can help farmers in decision-making from the initial stage (sowing the seeds of a favorable crop) of an agricultural process with ancillary advice about the crop suitable for a particular type of soil type.

Our proposed model *The Farmer's Support System* overcomes the setbacks by providing a three-component system for collection, prediction, and imparting output as shown in Figure 7.2. The first component incorporates a variety of sensors for collecting real-time data about the soil and surroundings using IoT technology. This collected data is sent to the database, which is further used by the second component. The second component concerns the prediction by means of an efficient ML algorithm to check and predict the success or failure of the particular crop with respect to the selected soil type. This predicted output is further sent for depiction through

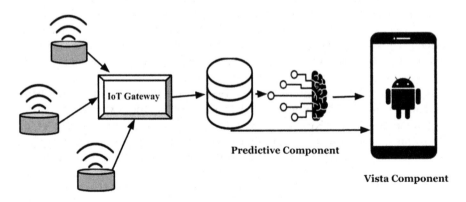

Figure 7.2   Three components of The Farmer's Support System.

a third component which has a user friendly, easy to understand user interface. The third component deals with effectively showing the output of prediction with auxiliary details, which can be helpful in better decision-making by the farmer.

The Farmers Support System works by the following three components:

- Tactile Component
- Predictive Component
- Vista Component

7.4.1 TACTILE COMPONENT

The *Tactile Component* comprises all the sensors used to find soil moisture, rainfall, moisture, humidity, temperature, and light intensity using IoT technology, where the data is collected by a preprogrammed microcontroller and sent to the database through a local server without human interference. IoT makes the dumb devices smarter to communicate with users and other devices on its own without human interference every single time. By means of IoT, the physical world can be connected to the digital world, which opens a path where imaginations are now turning into realities.

This component is made up of all the tangible elements of the system. As depicted in Figure 7.3, the data about the surroundings are accumulated through various sensors such as temperature, humidity, rainfall, light intensity, and soil moisture. These values are essential for prediction processing through an algorithm in further component, as the data collected by the sensors are primary attributes in the dataset used for successful prediction of the crop as shown in the description of the predictive component. A snapshot of the data collected from sensors is shown in Figure 7.4.

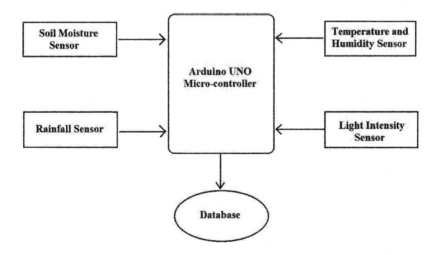

Figure 7.3 Flow diagram for sensors with microcontroller (MCU).

	A	B	C	D	E	F	G	H	I	J	K	L
1	Moisture	rainfall	Average Hu	Mean Temp	max Temp	Min temp	alkaline	sandy	chalky	clay	millet yield	Outcomes
2	12.80168	0.012361	57	62	71	52	0	1	0	0	2	1
3	12.85165	0.004172	57	58	73	43	0	1	0	0	0	1
4	12.77677	0	56	58	69	46	0	0	1	0	4	0
5	12.942	0.031747	62	56	70	43	0	1	0	0	0	1
6	12.98465	0	65	56	70	42	0	0	0	1	1	0
7	12.96447	0.027191	65	58	70	46	1	0	0	0	0	0
8	12.738	0.026821	61	56	70	42	0	0	0	1	1	0
9	12.81938	0.010284	58	57	72	42	0	0	0	1	1	0
10	12.88391	0.020465	63	60	76	45	0	0	1	0	0	0
11	12.78451	0.060054	62	59	71	47	0	1	0	0	4	1
12	12.96881	0.084119	56	58	69	46	0	1	0	0	2	1
13	12.78436	0	63	56	70	42	1	0	0	0	4	0
14	12.94459	0	67	58	72	43	0	1	0	0	4	1
15	12.92528	0.124479	58	60	75	46	0	0	0	1	1	0
16	12.82107	0.074505	59	58	68	49	1	0	0	0	0	0
17	12.93922	0.098584	53	60	68	53	0	0	0	1	1	0

Figure 7.4   A snippet of the selected dataset.

7.4.2   PREDICTIVE COMPONENT

The *Predictive Component* deals with the prediction of success or failure of the yield depending on soil type utilizing the information provided by the tactile component and a few user inputs. In addition to this, it also provides a list of crops with higher probability of success on the selected soil type.

The development process of *The Farmers Support System* saw a number of tasks requiring data analysis. Looking for a suitable dataset for applying a DL model and discarding the data sets that didn't have the required attributes. Furthermore, upon finalizing a data set, analysis of the relationships between the various attributes is performed, and if a primary relationship between the output and an attribute that isn't used as an input was found, it was discarded. Only after a data set has passed these processes of analysis, it is finalized to move forward to apply the algorithms on it.

7.4.2.1   Dataset Used

The dataset collected through various sensors in the first step has the following attributes:

- **Soil moisture:** This attribute gives the moisture level in the soil.
- **Rainfall:** This attribute gives the rainfall level on the soil.
- **Average humidity:** It gives the average humidity level.
- **Mean temperature:** It gives the average temperature of the soil.
- **Max temperature:** The most extreme temperature recorded for the soil is given by this trait.
- **Min temperature:** The minimum temperature recorded for the soil is provided by this attribute.
- **Alkaline:** This trait gives an estimation of 1 in the event of antacid soil and 0 in any case.
- **Sandy:** This trait gives an estimation of 1 if there is an occurrence of sandy soil and 0 in any case.

- **Chalky:** This property gives an estimation of 1 if is there an occurrence of chalky soil and 0 in any case.
- **Clay:** This property gives an estimation of 1 if there is an occurrence of clayey soil and 0 in any case.
- **Millet yield:** It gives the amount of millet yield.
- **Outcomes:** This property gives an estimation of 1 if there is an occurrence of harvest yield and 0 in any case.

Figure 7.4 represents the snapshot of the data set selected after data analysis and used by the predictive component for predicting the success or failure of the crop yield.

7.4.3 VISTA COMPONENT

The Vista component deals with the user interface and view layer for easy and better understanding of farmers about the environmental attributes required during crop production through android application.

Data that are expected as input come through IoT sensors and are effectively stored in the Google firebase, an action by Google which provides you the tools to produce high-quality apps. Real-time syncing makes it simple for users to the clients to recuperate their information from any gadget, be it web or versatile, while also helping the users to collaborate.

At the point when the clients go disconnected, the Realtime Database SDKs utilize nearby reserves on the gadget to help and store remodels. The local data are automatically synchronized with the device, whenever the device comes online. The real-time database can also integrate with firebase authentication to give a mere and programmed validation process.

Some data are also taken as user input such as type of soil and the name of the crop for the ease in predicting the crop. Also there is efficient implementation of the ML kit custom model feature of firebase for using the TensorFlow lite model in the android application. TensorFlow Lite is intended to make it accessible to perform ML on devices, instead of sending data back and forth from a server. By hosting the model with Firebase, ML Kit automatically updates the users with the latest version.

After hosting and configuring the proposed model, the input is provided by the user, and then it is sent to the model. The model will make all its predictions and shows the output on the screen of the phone indicating the success or the failure of the crop. In addition to this, it will also provide a list of suitable crops with respect to the selected soil type. Ancillary information about weather and soil conditions is also provided through the Vista component. Figure 7.5 depicts the functional requirements of the system keeping the needs of the targeted users in mind. It demonstrates the interaction of the users with the system.

7.4.4 TEST CASES DESIGN

Table 7.1 depicts how the system will work, albeit different situations when the end user is asked to provide the soil type as an input. The first case is when the soil type

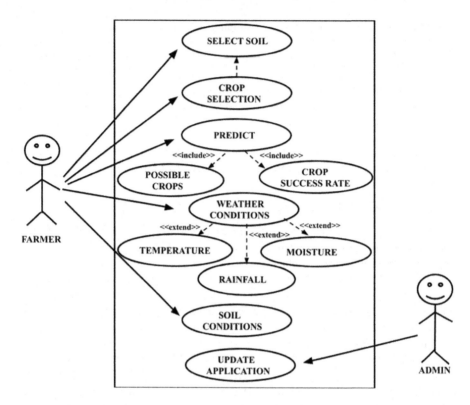

Figure 7.5 Use-case diagram for The Farmer's Support System UI.

Table 7.1
Test Cases for Get Prediction in the Farmer's Support System

Sl. No.	1	2
Test case name	select_soil_blank	soil_selected
Test procedure	Type of soil is not selected and Get Prediction is clicked.	Type of soil is selected and then the Get Prediction button is clicked.
Pre condition	The Get Prediction button is active.	The Get Prediction button is active.
Expected result	An error message must be displayed and the user should be prompted to select the type of soil (negative test case).	Proceed successfully (a new page is shown).
Reference to detailed design	Get Prediction module	Get Prediction module

Table 7.2

Testing for Get Prediction in the Farmer's Support System

Sl. No.	1	2
Test Case Name	select_soil_blank	soil_selected
Test Procedure	Type of soil is not selected and Get Prediction is clicked.	Type of soil is selected and then the Get Prediction button is clicked.
Pre Condition	The Get Prediction button is active.	The Get Prediction button is active.
Expected Result	An error message must be displayed and the user should be prompted to select the type of soil (negative test case).	Proceed successfully (a new page is shown).
Actual Result	Error message was displayed	Proceeded successfully to new page.
Remarks	Successful	Successful

text field is empty as the user has not provided any data but tries to get the result by clicking on the Predict button. The second case is when the user provides the soil-type input and clicks on the Predict button to get the required output.

7.4.5 SYSTEM TESTING

The objective of system testing is to assess end-to-end system specifications. Table 7.2 shows the tabular representation of the results acquired for the two test cases as mentioned in Table 7.1. The first one exhibits the scenario when the soil-type input is not provided, and the user tries to extract the output, whereas the second represents a scenario when the user provides soil type and proceeds with the process to get the results.

7.5 RESULTS AND DISCUSSION

The system is designed and built to provide an easy to use application to farmers in order to gain essential knowledge and information prior to sowing the crop seeds by incorporating technologies such as IoT, Data Analytic, Machine Learning, DL, and Android applications. The output attained can be categorized into three different categories.

7.5.1 OUTPUT FROM THE TACTILE COMPONENT

Figure 7.6 shows the different data fetched from IoT sensors which are used by the predictive component for further process.

```
10:54:52.835 -> =================================
10:54:52.835 -> Light: 117
10:54:52.835 -> Raining?: YES    Moisture Level: 324
10:54:52.883 -> Sample OK:
10:54:52.883 -> Temperature 42 *C,
10:54:52.883 -> Humidity 38 %
10:54:54.860 -> =================================
10:54:54.860 -> Light: 579
10:54:54.860 -> Raining?: YES    Moisture Level: 324
10:54:54.899 -> Sample OK:
10:54:54.899 -> Temperature 42 *C,
10:54:54.899 -> Humidity 38 %
10:54:56.883 -> =================================
10:54:56.883 -> Light: 557
10:54:56.883 -> Raining?: YES    Moisture Level: 324
10:54:56.944 -> Sample OK:
10:54:56.944 -> Temperature 42 *C,
10:54:56.944 -> Humidity 39 %
10:54:58.920 -> =================================
10:54:58.920 -> Light: 119
10:54:58.920 -> Raining?: YES    Moisture Level: 324
10:54:58.965 -> Sample OK:
10:54:58.965 -> Temperature 42 *C,
10:54:58.965 -> Humidity 39 %
10:55:00.948 -> =================================
10:55:00.948 -> Light: 116
10:55:00.948 -> Raining?: YES    Moisture Level: 324
10:55:00.998 -> Sample OK:
10:55:00.998 -> Temperature 42 *C,
10:55:00.998 -> Humidity 39 %
10:55:02.967 -> =================================
```

Figure 7.6 The data fetched from the sensors by the microcontroller.

7.5.2 OUTPUT FROM THE PREDICTIVE COMPONENT

From the data set described in Figure 7.4, it is clear that the Farmer's Support System does not need all the attributes in order to predict the crop outcome. The inputs are moisture, rainfall, average humidity, mean temperature, alkaline, sandy, chalky, clay. Min temperature, max temperature, and millet yield have been dropped. Out of the inputs, moisture, rainfall, average humidity, and mean temperature are provided through sensors in the soil and the type of soil-alkaline, sandy, chalky, or clay, is given by the farmer using the app through GUI. The output i.e. what is predicted is outcomes, giving a value of 1 in case of crop yield and 0 otherwise.

After performing a comparative study on various models such as regression, classification, DL, it was found that the accuracy and performance of DL models was higher than the others. Thus, moving forward with the ML algorithm, a DL algorithm is incorporated that deals with artificial neural networks.

The DL model that has been used is convolutional neural network(CNN). CNN is a type of feedforward artificial neural network and has various hidden layers. Usage of Sequential API of Keras is encompassed as it allows us to create models layer-by-layer, but it does not allow us to link multiple inputs, outputs layer. The model is also appropriate for a plain layers where each layer has exactly one input tensor and one output tensor.

Hidden layers of CNN are convulsed with the multiplication or dot product. 1D convolution layer is also called temporal convolution. This layer makes a convolution part that is convolved with the layer contribution over a solitary spatial measurement to create a tensor of yields. On the off chance that use_bias is True, at that point a predisposition vector is made and yield is included. On the off chance that enactment isn't None, at that point it is applied to the yields as well. When this layer is utilized as the main layer in a model, it provides an input_shape argument tuple of integers or None.

The activation function is ReLU as it removes negative values to avoid the problem of zero in summing up. Then, we have normalization after that final convolutional layer. Compile function is used to find accuracy and in this Adam as Optimizer is being used to reduce loss function.

The CNN is trained using the training set. Then, this trained model is used for making predictions on the input data being received. The model is further uploaded to the firebase using TensorFlow Lite so that it can be deployed in an android application. Figures 7.7–7.9 belong to the predictive component of the system illustrating the summary of the ML model applied , accuracy of the prediction done on the data set and a visual representation of the accuracy attained by means of a graph. The predicted output, which is shown in Figure 7.9, is then sent to the android app for displaying on the screen, thus helping the farmer find out the condition of the soil.

7.5.3 OUTPUT FROM THE VISTA COMPONENT

Figures 7.10–7.13, 7.14 and 7.15 pertain to the vista component of the system depicting the User Interface page in the android application which will be used by the farmers. The output that is provided to the end-user through android smartphones is shown in Figures 7.12, 7.13, 7.14, and 7.15, respectively.

The system is based on Firebase, which will have a predefined dataset present to be used for prediction. Data is sent through the tactile component installed in the farmer's plot of land along with the required input from the farmer through the Farmer's Support Application. For this purpose, IoT sensors are used, where without human interference the data are collected by the preprogrammed microcontroller and sent to the Firebase through a local server. Further, the prediction is done using CNN,

```
Layer (type)                    Output Shape           Param #
================================================================
conv1d (Conv1D)                 (None, 6, 32)           96

batch_normalization_v1 (Batc    (None, 6, 32)           128

dropout (Dropout)               (None, 6, 32)           0

conv1d_1 (Conv1D)               (None, 5, 64)           4160

batch_normalization_v1_1 (Ba    (None, 5, 64)           256

dropout_1 (Dropout)             (None, 5, 64)           0

flatten (Flatten)               (None, 320)             0

dense (Dense)                   (None, 64)              20544

dropout_2 (Dropout)             (None, 64)              0

dense_1 (Dense)                 (None, 1)               65
================================================================
Total params: 25,249
Trainable params: 25,057
Non-trainable params: 192
```

Figure 7.7 The data fetched from the sensors by the microcontroller.

```
Epoch 39/50
640/640 [==============================] - 0s 277us/sample - loss: 0.0992 - acc: 0.9719 - val_loss: 0.0256 - val_acc: 1.0000
Epoch 40/50
640/640 [==============================] - 0s 256us/sample - loss: 0.0832 - acc: 0.9828 - val_loss: 0.0239 - val_acc: 1.0000
Epoch 41/50
640/640 [==============================] - 0s 324us/sample - loss: 0.0891 - acc: 0.9828 - val_loss: 0.0222 - val_acc: 1.0000
Epoch 42/50
640/640 [==============================] - 0s 292us/sample - loss: 0.0784 - acc: 0.9875 - val_loss: 0.0206 - val_acc: 1.0000
Epoch 43/50
640/640 [==============================] - 0s 293us/sample - loss: 0.0661 - acc: 0.9906 - val_loss: 0.0192 - val_acc: 1.0000
Epoch 44/50
640/640 [==============================] - 0s 292us/sample - loss: 0.0685 - acc: 0.9937 - val_loss: 0.0178 - val_acc: 1.0000
Epoch 45/50
640/640 [==============================] - 0s 292us/sample - loss: 0.0750 - acc: 0.9766 - val_loss: 0.0167 - val_acc: 1.0000
Epoch 46/50
640/640 [==============================] - 0s 307us/sample - loss: 0.0581 - acc: 0.9937 - val_loss: 0.0156 - val_acc: 1.0000
Epoch 47/50
640/640 [==============================] - 0s 277us/sample - loss: 0.0573 - acc: 0.9937 - val_loss: 0.0146 - val_acc: 1.0000
Epoch 48/50
640/640 [==============================] - 0s 293us/sample - loss: 0.0570 - acc: 0.9906 - val_loss: 0.0137 - val_acc: 1.0000
Epoch 49/50
640/640 [==============================] - 0s 317us/sample - loss: 0.0547 - acc: 0.9953 - val_loss: 0.0129 - val_acc: 1.0000
Epoch 50/50
640/640 [==============================] - 0s 299us/sample - loss: 0.0703 - acc: 0.9828 - val_loss: 0.0121 - val_acc: 1.0000
[[0]]
```

Figure 7.8 Accuracy of the predictive model.

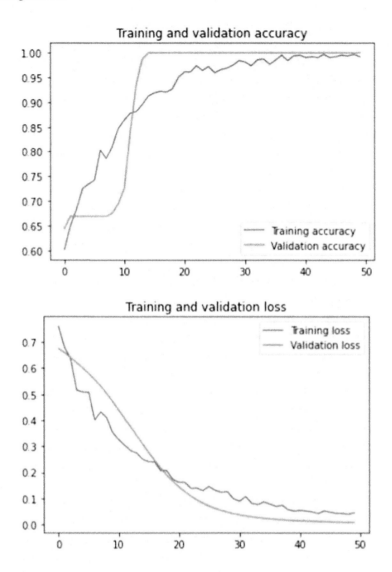

Figure 7.9 The accuracy graphs for predictive model.

which was selected after performing a comparative study on various models, and the output is generated. Finally, the output with a few ancillary details is displayed on the user Interface page of the Farmer's Support Application, for the ease and better understanding of the farmer.

The comfortable user experience is given at most priority in this system along with high accuracy of prediction.

Figure 7.10 The Front page outlook of the Farmer's Support application.

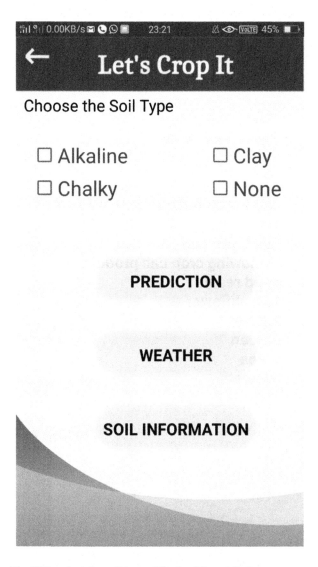

Figure 7.11 The UI to select the soil type of the land for prediction.

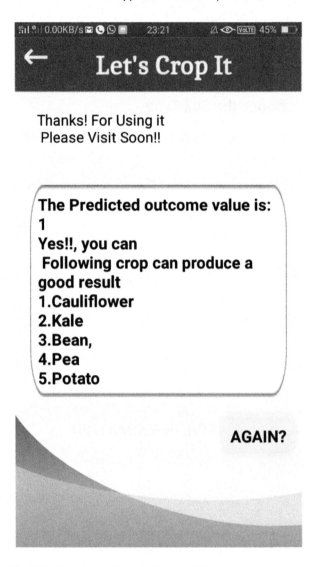

Figure 7.12   The UI if the output of the prediction will be a success.

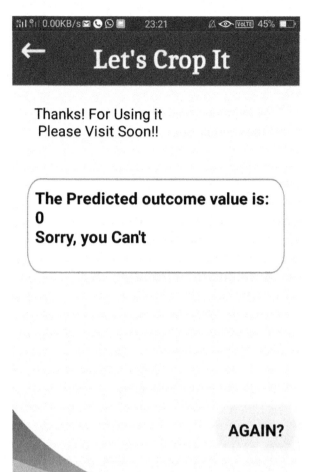

Figure 7.13 The UI if the output of the prediction will be not being success.

Thanks! For Using it
Please Visit Soon!!

TODAY'S WEATHER
1. **Temperature:- 45C**
2. **Humidity:- 28%**
3. **Moisture Level:-1015**
4. **Chances of Rainfall:- No**

Figure 7.14 The UI page displaying the weather conditions.

Figure 7.15 The UI page displaying the soil conditions.

7.6 CONCLUSION AND FUTURE SCOPE

The Farmer's Support System has been designed to decrease the number of farmers facing crop failure. Farmers are the backbone of our country, and if their condition improves, then it will surely be beneficial for the GDP of the country. It will also help farmers get more technologically inclined, and hence understand how science and technology can improve the methods of farming. Recently, not a lot of inventions have graced the world of agriculture and the Farmer's Support System hopes to start a series of inventions in this sector, thereby reducing crop failure rates and hoping to reduce cases of poverty-stricken farming families and farmer suicide rates.

The Farmer's Support System has a huge potential to be improved. In future, instead of the mobile application, a website could be developed. The website can also host platforms to get in touch with agriculture-based NGOs and help desks to get information about government schemes for farmers, discounts on seeds, and so on. It can be customized based on specific language and take aural inputs and provide auditory outputs. This project can be worked on and turned into a huge online community for farmers helping them every step of the way. In short, this project can be transformed into a massive resource for the agriculture community, helping the nation strengthen its primary sector.

REFERENCES

Aspelin, B., Kauppila, R., Kleemola, J., & Peltonen, J. (2001). Method for fertilizing cultivated plants for optimizing the quantity and quality of the yield. *Google Patents*.

Devika[1], B., & Ananthi, B. (2018). Analysis of crop yield prediction using data mining technique to predict annual yield of major crops.

Dumont, B., Basso, B., Bodson, B., Destain, J. P., & Destain, M. F. (2015). Climatic risk assessment to improve nitrogen fertilisation recommendations: A strategic crop model-based approach. *European Journal of Agronomy, 65*, 10–17.

Economic Survey *https://www.prsindia.org/report-summaries/economic-survey-2019-20*, Accessed: 2020-06-03.

Kaur, K. (2016). Machine learning: Applications in Indian agriculture. *International Journal of Advanced Research in Computer and Communication Engineering, 5(4)*, 342–344.

Kumar, R., Singh, M. P., Kumar, P., & Singh, J. P. (2015). Crop Selection Method to maximize crop yield rate using machine learning technique. In *2015 International Conference on Smart Technologies and Management for Computing, Communication, Controls, Energy and Materials (ICSTM), IEEE*, 138–145.

Liakos, K. G., Busato, P., Moshou, D., Pearson, S., & Bochtis, D. (2018). Machine learning in agriculture: A review. *Sensors, 18(8)*, 2674.

Limbore, N. V., & Khillare, S. K. (2015). An analytical study of Indian agriculture crop production and export with reference to wheat. *Review of Research Journal, 4(6)*, 1–8.

Mishra, S., Paygude, P., Chaudhary, S., & Idate, S. (2018). Use of data mining in crop yield prediction. In *2018 2nd International Conference on Inventive Systems and Control (ICISC), IEEE*, 796–802.

Nagini, S., Kanth, T. R., & Kiranmayee, B. V. (2016). Agriculture yield prediction using predictive analytic techniques. In *2016 2nd IEEE International Conference on Contemporary Computing and Informatics (ic3i)*, 783–788.

Nigam, A., Garg, S., Agrawal, A., & Agrawal, P. (2019). Crop yield prediction using machine learning algorithms. In *2019 Fifth International Conference on Image Information Processing (ICIIP), IEEE*, 125–130.

Patil, V. C., Maru, A., Shashidhara, G. B., & Shanwad, U. K. (2002). Remote sensing, geographical information system and precision farming in India: Opportunities and challenges. In *Proceedings of the Third Asian Conference for Information Technology in Agriculture*, 26–28.

Pongnumkul, S., Chaovalit, P., & Surasvadi, N. (2015). Applications of smartphone-based sensors in agriculture: A systematic review of research. *Journal of Sensors, 2015*, 18.

Priya, P., Muthaiah, U., & Balamurugan, M. (2018). Predicting yield of the crop using machine learning algorithm. *International Journal of Engineering Sciences & Research Technology, 7(1)*, 1–7.

Ramesh, D., & Vardhan, B. V. (2015). Analysis of crop yield prediction using data mining techniques. *International Journal of Research in Engineering and Technology, 4(1)*, 47–473.

Rajeswari, S. R., Khunteta, P., Kumar, S., Singh, A. R., & Pandey, V. (2018). Smart farming prediction using machine learning. *International Journal of Innovative Technology and Exploring Engineering (IJITEE), 8(7)*, 190–194.

Rajeswari, S.R., Khunteta, P., Kumar, S., Singh, A. R. & Pandey, V. (2019). Estimation of major agricultural crop with effective yield prediction using data mining. *International Journal of Innovative Technology and Exploring Engineering (IJITEE), 8(7)*, 170–174.

Salpekar, H. (2019). Design and implementation of mobile application for crop yield prediction using machine learning. In *2019 Global Conference for Advancement in Technology (GCAT), IEEE*, 1–6.

Schildroth, R., & Starr, D. B. (2018). Agronomic systems, methods and apparatuses. *U.S. Patent No. 10,028,426; issued July 24, 2018*.

Shakoor, M. T., Rahman, K., Rayta, S. N., & Chakrabarty, A. (2017). Agricultural production output prediction using supervised machine learning techniques. In *2017 1st International Conference on Next Generation Computing Applications (NextComp), IEEEE*, 182–187.

Singh, R. P., Prasad, A. K., Tare, V., & Kafatos, M. (2010). Crop yield prediction using piecewise linear regression with a break point and weather and agricultural parameters, *U.S. Patent No. 7,702,597*.

8 Development of Intelligent Internet of Things (IoT)-Based System for Smart Agriculture

Santosh Kumar Nanda, Archana Suresh, and Quilo Soman
Techversant Infotech Pvt Ltd
Trivandrum, India

Debi Prasad Tripathy
National Institute of Technology
Rourkela, India

Niranjan Ray
School of Computer Engineering, KIIT University
Bhubaneswar, India,

CONTENTS

8.1 Introduction 144
    8.1.1 Current Advance Applications of IoT 145
    8.1.2 Smart City 146
    8.1.3 Image Recognition 147
    8.1.4 Smart Homes 148
    8.1.5 Smart Energy 148
    8.1.6 Smart Traffic Control 149
    8.1.7 Smart Public Transportation 149
    8.1.8 Smart Healthcare 149
    8.1.9 Smart Farming (Agro Industry) 149
8.2 Internet of Things (IoT) Framework 150
    8.2.1 Amazon AWS: Internet of Things (IoT) Framework 150
    8.2.2 Microsoft Azure: Internet of Things (IoT) Framework 150
8.3 Need of Machine Learning in IoT Application 152
    8.3.1 Introduction to Machine Learning 152
8.4 Case Study: Smart Agriculture 155

8.4.1   Development of Regression Model for Rainfall Rate........................158
8.4.2   Development of Classification Model for Predicting
         Climate Condition...159
8.4.3   Development of Machine Learning Model for
         Recommending Crop 160
8.5 Summary..161
References...161

8.1 INTRODUCTION

The Internet of Things (IoT) is now helping all the mankind in many areas like health care, agriculture, smart home, smart city, education, entertainment industries, manufacture, mining, construction, defence, and so on. With advanced AI algorithm and advanced sensor management tools, IoT-based embedded system now helping all in everyday activities. Some major challenges associated with development process of IoT are highlighted as follows: (Ammar, Russello, & Crispo, 2017);

- Complex architecture of distributed and high-performance computing,
- Missing of proper methods and framework that reduce the error to low-level communication and easy installation of high-level architecture
- Different accessible of computing programming languages and
- Setup of robust communication environments

The abovementioned challenges help researchers, software and hardware engineers and developers to handle the computing infrastructure for all layers with both functional and non-functional requirements. These challenges also help to build IoT systems with less complexity and are able to maintain a robust computing balance environment. Figure 8.1 represents the demand of IoT in many areas (Singh, Tripathi, & Jara, 2014) Out of several applications of IoT, nowadays maximum applications are required for smart cities, smart homes and smart energy. The sudden technological improvement in IoT systems from integrated sensors to advanced cloud servers and

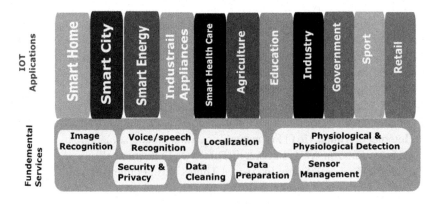

Figure 8.1   Application of Internet of Things in many areas.

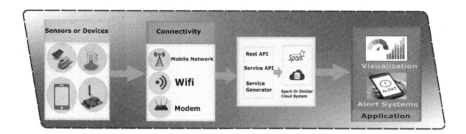

Figure 8.2 A low level system framework of IoT architecture.

storage defines the current demand for IoT systems. The prime objective of all IoT systems is to develop suitable and powerful connectivity between the Internet and sensors as well as to extract and store the data(Derhamy, Eliasson, Delsing, & Priller, 2015). With recent developments in IoT systems, it is possible to reduce human involvement, enrich the connection with embedded systems, and extract data from sensors in very faster way in all real-time environments. In this way, the IoT system is able to transform these high-volume data in a smarter way and with the current high-speed Internet protocol, and it is now possible to build smart applications. With powerful computing programming environments such as Python, Scala and Spark as well as big data database such as Hadoop, Amazon AWS and Microsoft Azure, it is now possible to design high volume data-based applications(Vijai & Sivakumar, 2016). It indicates that the scope of the IoT framework now expanding across all domains.

In the IoT system, there are two types of architecture available. Figure 8.2 represents low-level IoT system architecture, where Figure 8.3 represents the high-level IoT system architecture. In general, low-level IoT systems are developed with sensors and local network connectors only. The system has external data bases and has less complex computing environment. Low-level IoT system architecture is unable to maintain big data and also unable to connect with the outer world in a faster way. A low-level IoT system has connectivity with a hub, router, or modem and has the ability to send limited data (Petrolo, Loscrí, & Mitton, 2015) where high-level IoT system is the opposite to the low-level IoT system architecture. The big advantage of a high-level IoT system is the connectivity with cloud servers and cloud data storage devices. Therefore, cloud server and data storage are the backbone of high-level IoT system architectures, and it is also possible to establish customer data security in cloud (von Hippel E., 2005).

8.1.1 CURRENT ADVANCE APPLICATIONS OF IoT

IoT is getting accepted as an emerging technology worldwide. As reports mentioned, the number of Internet-connected devices has exceeded the human population of the world, and these IoT devices act as the main functional key for smart cities. However, IoT is now actively used in agro industrial and environmental problems (presented in Table 8.1). Figures 8.4 and 8.5 represent the demand for IoT applications in smart homes and smart cities (IoT market predictions and trends for 2018, n.d.).

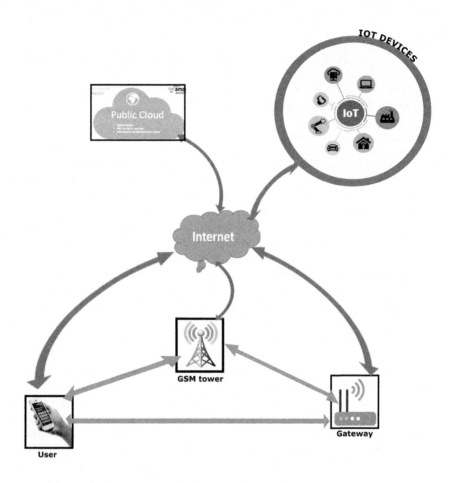

Figure 8.3 A high level system framework of IoT architecture.

8.1.2 SMART CITY

In every city, everyone gets services and has an opportunity to grow and avail a quality life, and therefore, the population is also high in cities. In the last 10 years (Figure 8.5), population has increased in cities compared to rural areas. Therefore, it is very necessary to establish better roads, communication and established better and advanced infrastructure to handle urbanization problems. IoT is now helping governing bodies to build smart cities by alarming unexpected problems like volcanos, earthquakes and providing better data analytics reports. Smart city requirements are currently limited, but demand will increase in the coming years (Mohammadi, Al-Fuqaha, Sorour, & Guizani, 2018). Smart cities use IoT where IoT can establish the

Table 8.1

General Application of IoT

IoT Application	Data Types	Data Processing
Smart health	Stream/Massive data	Edge/Cloud
Smart environment	Stream/Massive data	Cloud
Prediction	Stream data	Edge
Smart citizen	Stream data	Cloud
Smart agriculture	Stream data	Edge/Cloud
Smart home	Massive/Historical data	Cloud
Smart air controlling	Massive/Historical data	Cloud
Smart public place	Massive/Historical data	Cloud

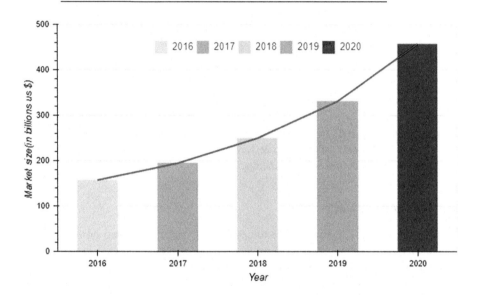

Figure 8.4   Market size for IoT applications.

connectivity between the customers using advanced sensors (RFID, GPS, IR, etc.). IoT system is capable of transforming and gathering sensor data as well as using the data and using machine learning algorithm to predict future requirements. With the advanced support from IoT, smart cities can now develop with low-cost budgets(Mahdavinejad et al., 2018).

8.1.3   IMAGE RECOGNITION

In recent years, mobile technology has advanced significantly, and now all mobile devices are equipped with high-resolution camera. Therefore, everybody wants to protect these devices, and with help of advanced artificial intelligence tools like deep

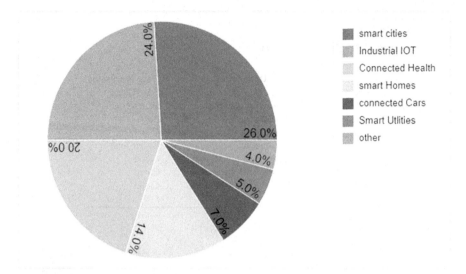

Figure 8.5   Global IoT market share by subsector.

learning (DL) and IoT, it is possible to develop image recognition-based security for all the smartphone users (Bach & Jordan, 2002). This image recognition-based security is also needed in smart home appliances.

8.1.4   SMART HOMES

With the help of IoT, it is now able to develop the concept of smart home and able to enhance the personal security. Now it is also possible to integrate high artificial intelligence-based algorithm in IoT to predict the health status, faster control in the home appliances, helping family to maintain a better security with advanced face recognition tools. Modern IoT system now integrates with cloud server and storage system and with high-speed Internet protocol, IoT system enables the remote monitoring and very robust management of appliances and systems. Therefore, now smart home monitors family members' activities and the internal environment at home, and this way, it provides better services that fulfil the exact and demands of a user (Mohammadi et al., 2018).

8.1.5   SMART ENERGY

In recent years, smart energy or smart energy management has been an important research subject for all the researchers and energy producing industries. Every developing and developed country is now focusing on smart energy projects to fulfil the power demand and increase safety. Therefore, IoT is now very essential in smart energy projects to maintain environmental clearance and safety. With high research development in wireless communication, conversation of 3G to 4G LTE and advanced 5G networks now every energy industry is planning for big milestones in the next 5 years. It is now possible to use renewable sources with IoT to reduce the power consumption and can predict the user's exact power requirement with respect to the users location and financial condition (Mahdavinejad et al., 2018).

8.1.6 SMART TRAFFIC CONTROL

To maintain advanced safety management in traffic system, IoT-based system is required. The IoT system can receive and transfer sensor data used in traffic system, and using advanced data analytics, it is now possible to predict any unwanted issues. It is also possible to design DL-based computer vision solution using road camera data. The IoT system now enables road safety in real-time environment (Mahdavinejad et al., 2018).

8.1.7 SMART PUBLIC TRANSPORTATION

Smart public transportation is now very essential in smart city projects and also a part of intelligent transportation system. It is now possible to design better, smarter and safer smart transportation system using IoT and Artificial Intelligence algorithm. Quick development in sensor management, equipment and computing devices helps IoT-based public transportation system to enrich in public safety, comfortability, and able to reduce the accident rate. Another challenge in smart transportation system is to handle traffic volume and reduce the pollutant parameters due to high traffic volume. Therefore, advanced IoT system provides better opportunity in smart public transportation system to optimize the above two major challenges (Mohammadi et al., 2018).

8.1.8 SMART HEALTHCARE

Health care is one of the important sectors of business in every developing country. Now, IoT helps to develop a better relationship between doctors and patients and indirectly helps to increase the revenue. With the available advanced sensors and high-speed Internet connectivity, it is now easy to transfer patient's critical data and possible to enhance the patient's health status. IoT is now helping health-care industry to set up excellent virtual consulting system with less error and also able to manage patient's health status. With advanced artificial intelligence algorithm and advanced computer vision algorithm, IoT helps to predict patient's future health condition and able to improve the quality of patient's personal life (Senders et al., 2017).

8.1.9 SMART FARMING (AGRO INDUSTRY)

Increased population is now a big issue everywhere, and problems such as environmental changes, climate change, unpredictable weather condition are now additional issues with population (Talavera et al., 2017). To fulfil the demand, now every agricultural industry, farmers need high edge technology in farming, and IoT helps a lot in this sector (Yang et al., 2017). Smart agriculture is now integrated with an IoT-based embedded system that can predict temperature and humidity conditions in advance and able to provide recommended seasonal crop to farmer and agricultural industries (Mishra et al., 2019).

In this chapter, we presented a case study of smart framing in detail. In this chapter, different types of IoT frameworks are discussed in Section 8.2. Section 8.3 provides a brief introduction to machine learning algorithm and the integration process with IoT devices. The case study of smart agriculture/framing is represented in Section 8.4. Conclusion is presented in Section 8.5.

8.2   INTERNET OF THINGS (IoT) FRAMEWORK

With increased application of IoT, many IoT framework applications have been developed. Using these frameworks, now IoT application architectures are possible to design (Shah, Garg, Sisodiya, Dube, & Sharma, 2018). There are some good IoT framework available :AWS IoT from Amazon, ARM Bed from ARM and other partners, Azure IoT Suite from Microsoft, Brillo/Weave from Google, Calvin from Ericsson, HomeKit from Apple, Kura from Eclipse, and Smart Things from Samsung. Out of these available frameworks, AWS and Azure IoT frameworks are widely used by following the criteria:

- Importance of service providers in the software, hardware-infrastructure and electronics industries,
- Recent research and product development and the amount of development robust application available on store,
- The market demand of IoT.

Here in this analysis only two popular frame work design discussed (Prathibha, Hongal, & Jyothi, 2017).

8.2.1   AMAZON AWS: INTERNET OF THINGS (IoT) FRAMEWORK

AWS (Amazon Web Services) IoT is a popular and widely used cloud platform released by Amazon. The framework results in the smart devices securely interacting and easily connecting with the AWS cloud and other devices which are connected to it. In general, with AWS IoT, it is easy for any user to understand how to use AWS with its variety of applications and database resources such as Amazon Dynamo DB, Amazon S3, Amazon Machine Learning and others. AWS IoT allows applications to communicate with all sensors, modems and all data related devices even when they are offline. Figures 8.6 and 8.7 represents the architecture of AWS IoT (Doshi, Patel, & kumar Bharti, 2019).

8.2.2   MICROSOFT AZURE: INTERNET OF THINGS (IoT) FRAMEWORK

Compared to the AWS IoT service, Azure IoT is a very robust and cost-effective application released by Microsoft. Microsoft Azure is also very user-friendly application platform that provides a set of services that enable end-users to interact

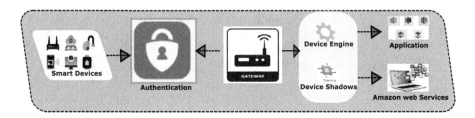

Figure 8.6   AWS IoT architecture (Amazon).

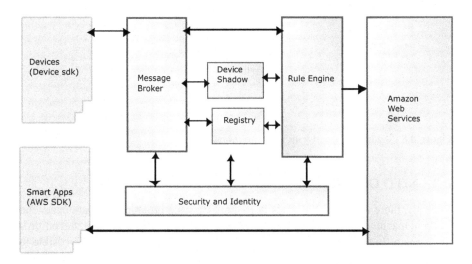

Figure 8.7   AWS IoT security mechanism.

with their IoT devices, receive and transfer data from them, perform various operations over data (e.g. data distribution, data analysis, transformation, etc.) and data
visualization in all types of user requirement. Azure IoT suite offers a full-featured
IoT framework as a combination of three different data associated problems are: big
data, telemetry patterns, and scaling. It supports programming languages, operating
systems, and wide range of hardware devices. Figure 8.8 represents the Azure IoT
architecture (Singh, Chaturvedi, & Akhter, 2019)

Various research sources estimate that the total number of Internet-connected instruments in use will be between 35 and 60 billion in the next 10 years, indicating
increased demand. Increased research in communication technology, e.g. 5 G network, now helps all telecom manufacture companies to expand the bandwidth and
also helps to set up an excellent communication between cloud services and IoT devices. Intelligent analysis and processing of this Big Data is the key to development
of smart IoT applications. For better performance of IoT application, particularly
in smart home or smart city application, machine learning tools are required (Ayaz,
Ammad-Uddin, Sharif, Mansour, & Aggoune, 2019).

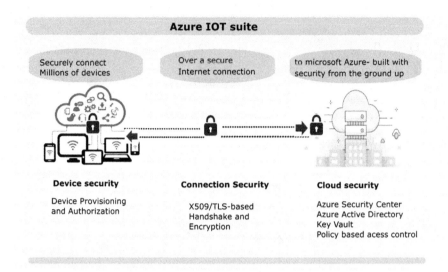

Figure 8.8   Azure IoT security architecture (Microsoft).

8.3   NEED OF MACHINE LEARNING IN IoT APPLICATION

Recently, there are a lot of updates in Python, R, Scala programming language, and so on in machine learning field and possible to integrate with IoT-integrated embedded system. In almost all IoT-based projects, sensors are used, and it is possible to extract and analyze data with available modern database tools such as monogo-db, cassandra, radis and so on. Therefore, machine learning algorithm like support vector machine (SVM), cluster algorithms like principle of component analysis (PCA), radial basis function (RBF) network, convolutional neural network (CNN), and so on now widely used in smart health care, smart city, smart farming to predict future requirements(Lecun, Bottou, Bengio, & Haffner, 1998). Small portable and advance IoT computing device like RaspberryPI-4 have now available with python support, and it is now possible to integrate machine learning algorithm in it(Nanda, Tripathy, Nayak, & Mohapatra, 2013).

8.3.1   INTRODUCTION TO MACHINE LEARNING

Machine Learning is a branch of artificial intelligence based on the principle that machines should be able to learn and receive knowledge and inference through long learning period. A lot of mathematical operations like classification, regression and optimization are associated with Machine learning methods. Figure 8.9 represents the difference between general machine learning and machine learning. With advanced computer programming language support like python, it is now possible to integrate machine learning algorithm with IoT devices and the principle of machine learning as follows: (Mahapatra, Nanda, & Panigrahy, 2011):

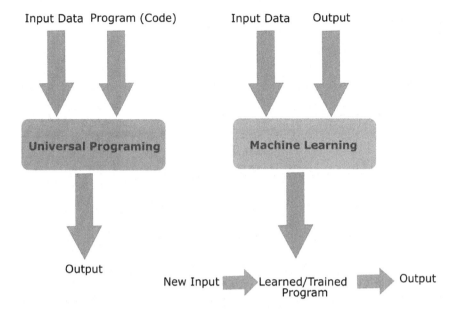

Figure 8.9 Differences between general programming and machine learning.

- The high demand for machine learning in important areas such as health-care, education, energy, telecom, fashion and other sectors.
- Product recommendation in every domain such as telecom, health care, re-tail, and so on.
- Customer sentiment analysis using customer reviews and social network posts.
- Fraud detection in banking domain is now essential demand of machine learning.

In general, machine learning is three types: supervised, reinforcement and unsuper-vised learning. The training set is a set of input taken as samples by the learning algo-rithm. Supervised, unsupervised, and reinforcement are the three main categories of learning. In supervised learning, each training data set contains the input vectors and with their corresponding target output vectors, which is known as labels. In unsuper-vised learning, labels are not a requirement for the training set. Reinforcement learn-ing always deals with the problem of getting the appropriate action or list of actions to be taken about a given situation in order to increase the performance. Figure 8.10 represents the structure of machine learning tools in brief (Mahapatra et al., 2011) The principle of supervised learning is to predict similar output corresponding to the target output with given historical input data.The performance of supervised learning increases with the volume of data set. Classification is required where target value is categorical and regression is the opposite to classification. In classification, target

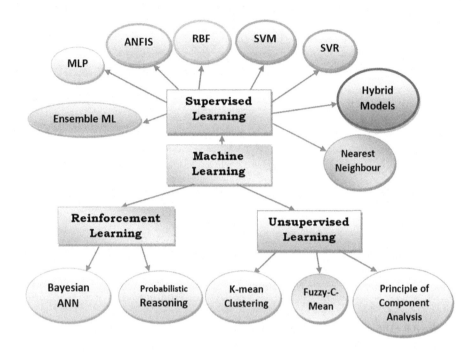

Figure 8.10 Details of machine learning tools.

values need to be converted to a finite value, whereas in regression, the real value of the target output must be used. In unsupervised learning, as target output is missing; therefore, it is very difficult to estimate the error. Therefore, it is a very challenging task to design unsupervised learning architecture. Clustering is one of the major sectors in unsupervised learning methods, where it has the principle of identifying the sensible clusters of similar samples within the input data. To increase the accuracy of any machine learning model, it is important to generate the exact feature of the input data set and this process is called feature engineering. Feature extraction is the major part of feature engineering that has a preprocessing stage. The performance of the machine learning algorithm will enhance and improve after the addition of the feature extraction step (Mahapatra et al., 2011). Some of the good applied machine learning algorithms for data analysis are tabulated in Table 8.2.

It is now possible to integrate advanced artificial intelligence algorithms like DL, computer vision algorithm in IoT devices and it is represented in Figure 8.11. DL has been actively utilized in many IoT applications in recent years and successfully work for many IoT projects like Smart City. Particularly, DL algorithm used for object detection task. Figure 8.11 represents the integration process of DL, high computing machine learning. Now it is possible to integrate IoT devices with Amazon or Azure IoT cloud and possible to receive and transfer data in real time.

Table 8.2

Overview of Frequently Used Machine Learning Algorithms for Smart Data Analysis

Machine Learning Algorithm	Data Processing Tasks
K-Nearest neighbors	Classification
Naive Bayes	Classification
Support vector machine	Classification
Linear regression	Regression
Support vector regression	Regression
Classification and regression trees	Classification/regression
Random forests (Ensemble)	Classification/regression
Bagging and stacking (Ensemble)	Classification/regression
K-Means	Clustering
Fuzzy-C-Mean	Clustering
Density-Based spatial clustering of applications with noise (DBSCAN)	Clustering
Principal component analysis (PCA)	Feature extraction
Canonical correlation analysis	Feature extraction
Feed forward neural network	Regression/Classification/ Clustering/Feature extraction
Radial basis function network	Regression/Classification/ Clustering/Feature extraction

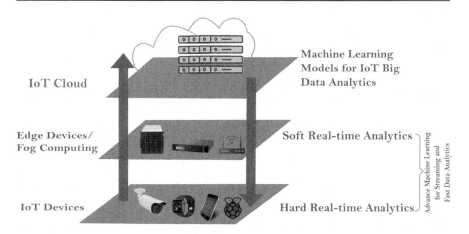

Figure 8.11   IoT data generation at different levels and deep learning models to address their knowledge abstraction.

8.4   CASE STUDY: SMART AGRICULTURE

Agriculture is a major economic source of our country. Biology, climate, economy and geography are various factors which play an important role in agricultural crop production. Different factors have different effects on agriculture. Crops are the basic necessity of life. Growing crops are important because they are a major source of

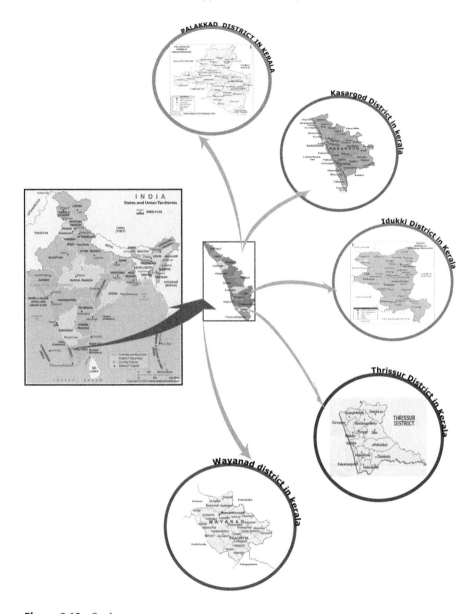

Figure 8.12 Study area.

food. Each crop is cultivated in different seasons, and the availability of the rain is the most important factor. Achieving maximum crop yield is the main objective of agricultural production. However due to some natural disasters, majority of farmers are not getting the expected crop yield.

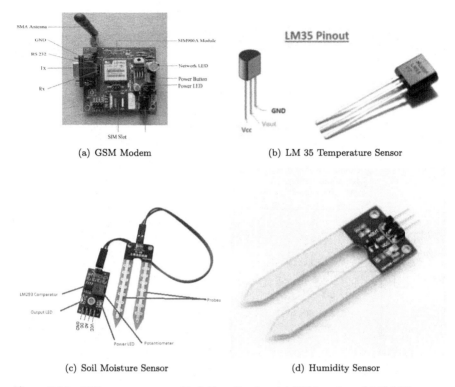

(a) GSM Modem

(b) LM 35 Temperature Sensor

(c) Soil Moisture Sensor

(d) Humidity Sensor

Figure 8.13   Different sensors used in IoT application: (a) GSM modem, (b) LM 35 temperature sensor, (c) soil moisture sensor and (d) humidity sensor.

To overcome these problems, build a recommended system of agriculture production and distribution for farmers. By which farmers can make decision in which month, which crop is suitable for sowing so that they can get more benefits. Here, only weather data like rainfall, temperature and relative humidity are considered. Based on these factors future rainfall rate can be predicted. After that, based on these weather conditions, predict the land that is suitable for cultivation. If the land is suitable for cultivation, the crops that are suitable at that time are recommended. For this case study, we selected five places in Kerala. Figure 8.12 shows the places that were chosen for the analysis (Palakkad, Idukki, Thrissur, Wayanad, and Kasargod).

In this chapter, a case study has been proposed. This application is a IoT-based smart agricultural application. Four types of sensor were used to collect data from the respective area. GSM modem, temperature sensor, soil moisture, and humidity sensor used for this IoT application. Figure 8.13 represents the detail about the sensors. Soil moisture sensor was used to get whether crop field need water supply or not. In this application, to recommended crop it is needed to know about the rainfall and it is very difficult to get the exact rainfall value and not even possible to establish a meteorological sensor which is very costly for farmers to implement. Therefore,

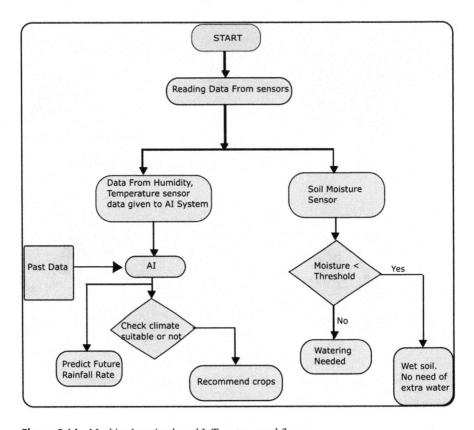

Figure 8.14   Machine learning based IoT system workflow.

machine-learning helps to predict exact rainfall rate by taking temperature and humidity data. As per Figure 8.14, machine learning models were used three times. The first machine learning model is a regression model is used to predict the rainfall rate by considering temperature and humidity data. The second machine learning model is used to predict whether or not the climate is suitable for cultivation and the last one is recommended respective crop as per the climate record.

8.4.1 DEVELOPMENT OF REGRESSION MODEL FOR RAINFALL RATE

For developing of a regression model, we need to develop data set with independent and dependent variables. Here the independent variables are the temperature and humidity and the dependent variable is rainfall rate. Three different models, namely, multilayer perceptron (MLP), logistic regression, and multi linear regression (MLR), are used to predict the rainfall data. For performance evaluation, R^2 values are considered. Table 8.3 represents the performance of different machine learning model results, and it is found MLR is good at predicting rainfall rate.

Table 8.3

Performance of Machine Learning Based Regression Model for Rainfall Rate for Different Locations

Month	Place	MLP-Regression	Logistic Regression	MLR
Jan	Palakkad	$R^2 = 0.85$	$R^2 = 0.83$	$R^2 = 0.92$
May	Idukki	$R^2 = 0.82$	$R^2 = 0.86$	$R^2 = 0.93$
August	Thrissur	$R^2 = 0.81$	$R^2 = 0.88$	$R^2 = 0.91$

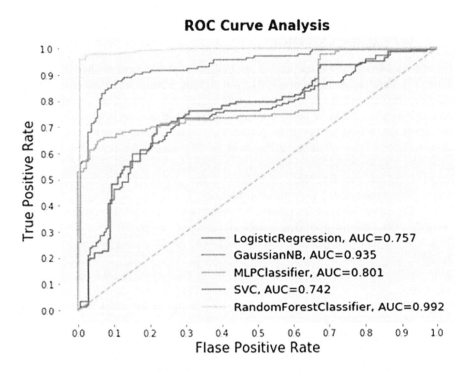

Figure 8.15 ROC curve comparison.

8.4.2 DEVELOPMENT OF CLASSIFICATION MODEL FOR PREDICTING CLIMATE CONDITION

After predicting the rainfall rate, it is used to know if the weather is good for cultivation or not. With available temperature and humidity and predicted rainfall rate, machine learning models were designed to predict the climate condition for the cultivation. Different classification models such as Naviee Bayes, SVM, ensemble machine learning model were used. Figure 8.15 represent the ROC comparison and Table 8.4 represent the comparison result between the applied machine learning models.

Table 8.4
Confusion Matrix

Model	Logistic	MLP	SVM	NB	Randomforest
TP	118	100	161	36	153
TN	55	52	100	110	59
FP	55	100	44	100	42
FN	113	171	118	178	160

8.4.3 DEVELOPMENT OF MACHINE LEARNING MODEL FOR RECOMMENDING CROP

For the recommendation, the KNN algorithm is used. Predict whether or not the climate is suitable based on predicted rainfall, temperature, and humidity; if it is suitable, recommend crops to farmers based on rainfall so that they can maximize productivity and earn more profit from agriculture. Crop data set is used for the recommendation. Based on the given parameters (rainfall, temperature), find out the crops suitable for this climate. For this work, KNN works as follows (Table 8.5):

Step 1: Load the crop data set.
Step 2: Select the number K of the neighbors.
Step 3: Calculate the Euclidean distance between each data in the crop data set and given data (that means given temperature and predicted rainfall for that place).
Step 4: As per the calculated Euclidean distance, the K-nearest neighbors are selected.
Step 5: From these k neighbors, the number of the data points in each categories are counted.
Step 6: The new data points to that category are assigned for which the number of the neighbors are maximum.
Step 7: Store all the calculated distance.
Step 8: Sort all distance.
Step 9: Pick the first three entries from the sorted collection.

Table 8.5
Model Result for Recommended Crop

Location	Temperature	Humidity	Predicted Rainfall Rate	Recommended Crops
Idukki	21	85	197.30	Coffee, Turmeric, Sugurcane
Palakkad	24	82	128.83	Sugarcane, cotton, millets
Thrissur	26	85	51.18	Gram, oilseeds, groundnut
Kasargod	27	80	214.5	Clove, Turmeric, coffee

8.5 SUMMARY

In this chapter, we discussed how IoT and machine learning helping farmers to set-up a smart agriculture with a low cost investment and we also represents recent application of IoT in various domain. Different application of IoT like smart city, smart health care, smart home and smart agriculture also discussed in this chapter. This case study presented in this chapter is a current IoT application in smart agriculture. With the use of different sensor and machine learning tool IoT device can able to recommended crops to farmers. This case study is now a prototype solution and can be commercialize in future. This research output helps researcher to adopt best machine learning tool for IoT application. The results of different models are compared based upon the R^2, root mean square, and percentage prediction error. The model which gives the lower root mean square, percentage prediction and higher R^2 static value is to be considered to be the best model for rainfall prediction.In this project, multiple regression is better for predicting rainfall. For classification, Random Forest gives more accuracy. This application is more helpful to farmers, they can maximize the productivity and earn more profit from agriculture.Also reduce the death rate of farmers.

REFERENCES

Ammar, M., Russello, G., & Crispo, B. (2017). Internet of things: A survey on the security of iot frameworks. *Journal of Information Security and Applications*, *38*, 8–27.

Ayaz, M., Ammad-Uddin, M., Sharif, Z., Mansour, A., & Aggoune, E. M. (2019). Internet-of-things (iot)-based smart agriculture: Toward making the fields talk. *IEEE Access*, *7*, 129551–129583.

Bach, F. R., & Jordan, M. I. (2002). Kernel independent component analysis. *Journal of Machine Learning Research*, *3*, 1–48.

Derhamy, H., Eliasson, J., Delsing, J., & Priller, P. (2015). A survey of commercial frameworks for the internet of things. In *2015 IEEE 20th Conference on Emerging Technologies and Factory Automation (etfa)* (pp. 1–8).

Doshi, J., Patel, T., & kumar Bharti, S. (2019). Smart farming using iot, a solution for optimally monitoring farming conditions. *Procedia Computer Science*, *160*, 746–751. Retrieved from http://www.sciencedirect.com/science/article/pii/S1877050919317168 (The 10th International Conference on Emerging Ubiquitous Systems and Pervasive Networks (EUSPN-2019) / The 9th International Conference on Current and Future Trends of Information and Communication Technologies in Healthcare (ICTH-2019) / Affiliated Work-shops)

IoT market predictions and trends for 2018. (n.d.). *Website:https://infotechlead.com/iot/iot-market-predictions-and-trends-for-2018-52448).*

Lecun, Y., Bottou, L., Bengio, Y., & Haffner, P. (1998). Gradient-based learning applied to document recognition. *Proceedings of the IEEE*, *86* (11), 2278–2324.

Mahapatra, S., Nanda, S. K., & Panigrahy, B. (2011). A cascaded fuzzy inference system for indian river water quality prediction. *Advances in Engineering Software*, *42* (10), 787–796. Retrieved from http://www.sciencedirect.com/science/article/pii/S0965997811001256

Mahdavinejad, M. S., Rezvan, M., Barekatain, M., Adibi, P., Barnaghi, P., & Sheth, A. P. (2018). Machine learning for internet of things data analysis: A survey. *Digital Communications and Networks*, *4*(3), 161–175. Retrieved from http://www.sciencedirect.com/science/article/pii/S235286481730247X

Mishra, D., Pande, T., Agrawal, K. K., Abbas, A., Pandey, A. K., & Yadav, R. S. (2019). Smart agriculture system using iot. In *Proceedings of the Third International Conference on Advanced Informatics for Computing Research*. New York, NY, USA: Association for Computing Machinery. Retrieved from https://doi.org/10.1145/3339311.3339350

Mohammadi, M., Al-Fuqaha, A., Sorour, S., & Guizani, M. (2018). Deep learning for iot big data and streaming analytics: A survey. *IEEE Communications Surveys Tutorials*, *20* (4), 2923–2960.

Nanda, S. K., Tripathy, D. P., Nayak, S. K., & Mohapatra, S. (2013). Prediction of rainfall in india using artificial neural network (ann) models. *International Journal of Intelligent Systems and Applications*, *5*, 1–22.

Petrolo, R., Loscrì, V., & Mitton, N. (2015). Towards a smart city based on cloud of things, a survey on the smart city vision and paradigms. *Transactions on Emerging Telecommunications Technologies*, 1–11. Retrieved from http://dx.doi.org/10.1002/ett.2931

Prathibha, S. R., Hongal, A., & Jyothi, M. P. (2017). Iot based monitoring system in smart agriculture. In *2017 International Conference on Recent Advances in Electronics and Communication Technology (ICRAECT)* (p. 81–84).

Senders, J. T., Staples, P. C., Karhade, A. V., Zaki, M. M., Gormley, W. B., Broekman, M. L. D., . . . Arnaout, O. (2017). Machine learning and neurosurgical outcome prediction: A systematic review. *World Neurosurgery*, 1–11.

Shah, U., Garg, S., Sisodiya, N., Dube, N., & Sharma, S. (2018). Rainfall prediction: Accuracy enhancement using machine learning and forecasting techniques. In *2018 Fifth International Conference on Parallel, Distributed and Grid Computing (PDGC)* (p. 776–782).

Singh, D., Tripathi, G., & Jara, A. (2014). A survey of internet-of-things: Future vision, architecture, challenges and services. In *2014 IEEE World Forum on Internet of Things (WF-IoT)* (pp. 287–292).

Singh, N., Chaturvedi, S., & Akhter, S. (2019). Weather forecasting using machine learning algorithm. In *2019 International Conference on Signal Processing and Communication (ICSC)* (p. 171–174).

Talavera, J. M., Tobón, L. E., Gómez, J. A., Culman, M. A., Aranda, J. M., Parra, D. T., . . . Garreta, L. E. (2017). Review of iot applications in agro-industrial and environmental fields. *Computers and Electronics in Agriculture*, *142*, 283–297. Retrieved from http://www.sciencedirect.com/science/article/pii/S0168169917304155

Vijai, P., & Sivakumar, P. B. (2016). Design of iot systems and analytics in the context of smart city initiatives in india. In *Procedia Computer Science* (Vol. 92, pp. 583–588).

von Hippel E. (2005). Democratizing innovation: The evolving phenomenon of user innovation. *Journal für Betriebswirtschaft*, *55*, 63–78.

Yang, Y., Sun, L., Hu, J., Porter, D., Marek, T., & Hillyer, C. (2017). A reliable soil moisture sensing methodology for agricultural irrigation. In *2017 IEEE International Symposium on Parallel and Distributed Processing with Applications and 2017 IEEE International Conference on Ubiquitous Computing and Communications (ISPA/IUCC)* (p. 1342–1349).

9 IoT in Health Care in the Context of COVID-19

An Overview on Design, Challenges, and Application

P. Ravichandran
Binary University
Puchong, Malaysia

K.K. Venkataraman
PSG Institute of Advanced Studies (PSGIAS)
Coimbatore, India

CONTENTS

9.1 Introduction 163
9.2 Design and Challenges 165
9.3 IoT Integration of Previous Generation Health-care Equipment................. 166
9.4 Digital Twin Concepts on Health care using IoT Platform 168
9.5 IoT and Mobile Applications in Health Care 170
9.6 IoT and Artificial Intelligence Application in Health care........................... 171
9.7 Sensors for IoT Application in Health care... 172
9.8 IoT and its Security Aspects... 173
9.9 Remote Monitoring using IoT in Health Care in the Context
    of COVID-19 174
9.10 Conclusion 175
Acknowledgment 175
References.. 176

9.1 INTRODUCTION

As electronics and related communication technologies are advancing in a fast phase, it is pushing designers toward re-engineering of the existing products as well as developing new products to embrace the new technologies like Internet of Things (IoT) and artificial intelligence (AI), and so on. IoT can be defined in simple terms: 'Data is no longer confined to the individual system or product or immediate environment but gets on to the outside world like Internet'. In technological terms, the IoT is a system of interrelated computing devices, mechanical and digital machines, sensors, objects, people that are provided with unique identifiers and the ability to transfer data over a network without requiring human-to-human or human-to-machine interaction.

Since health care is one of the critical aspects of society, the applications of new technologies in this domain cannot be ignored. The US National Academy of Engineering has identified the health care domain as one of the challenges and opportunities to do research and application in IoT and Cyber Physical Systems (Kyoung-Dae & Kumar, 2013). However, there needs to be a cautious approach in introducing new technologies in health care as it deals with the life of people, especially in invasive devices. Also, in the case of IoT, the data from patients or even healthy people on health is a sensitive issue and needs to be handled carefully and with utmost security. Currently, there is a huge gap between integrating data from various medical devices and manual maintenance of patient records in hospitals which can be bridged by IoT. Data on health care is one huge treasure which can be the foundation for improving the health aspects of society at large. The new technologies such as 5G, Cloud storage and services, and so on, provide huge support in faster handling large volumes of data, filtering/analyzing/ transforming it into useful information. With the boom in the mobile platforms, health care IoT on mobile applications is a further enhancement that could be used to monitor, diagnose, and prevent health care issues considering its mobility, seamlessness, and convenience (Yanxiang Guo et al., 2017).

Implementation of new technologies is fairly easier in non-invasive health care devices. IoT is very relevant in a contagious environment like COVID-19, wherein the proximity to patients to monitor their parameters is a big risk to the medical fraternity in terms of infections. The patient parameters can be monitored continuously without getting near the patients, through data transfer outside the product, through IoT technology. There are different approaches, platforms, and use cases in the IoT domain due to the surge in technology and application requirements (Jasmin Guth et al., 2018). Though there are a large number of ready-made hardware available for designs for a quick development in the IoT domain, yet it is advisable to have a dedicated design that has cost and security advantages. The software that goes into the IoT system should be modular, scalable, connected, and reliable (Christian Legare, 2014). With the advent of new technologies like 5G, Cloud storage and services, there is an increasing trend in the application of IoT in health care, both in terms of technological advances and social impact.

Further, a health-care IoT supports preventive medical care as well by accumulating data from daily checks on self-care medical devices such as blood pressure (BP) monitors, glucose monitors, and so on, which can analyze the day-to-day variations and alert the medical support system on the alarming trends. With health-care IoT, patients can have monitoring that ensures they get the right prescriptions and the encouragement to take medications on time, warnings of impending crises and less time in hospitals.

9.2   DESIGN AND CHALLENGES

The data transmission outside the product through IoT is very relevant in the current situation like a COVID-19 type of environment, where people need to be quarantined (either in ICU or in Quarantine wards or at home) and monitoring their vital parameters from a remote location is an essential part of the treatment to avoid infection to the medical fraternity.

A simple block diagram, as shown in Figure 9.1, depicts the application of IoT in health care. The product is connected to the Internet through different technologies like Wi-Fi, Bluetooth. Once the data is outside the product, one can store it in the cloud or in a central storage system. Integration between IoT and Cloud services as well as emerging complex applications requires a uniform software layer on top of these blended IoT elements and cloud services (Hong Linh Troung & Schahram Dustdar, 2015).

One of the major design challenges in an IoT environment is the power consumption for portable devices which are battery-powered and mobile devices (Yanxiang Guo et al., 2017). Complex designs in sensor signal processing, user interface, security, and wireless connectivity consume a lot of power and we need to adopt new strategies to overcome the situation. One of the solutions is to use the sleep/deep sleep capabilities of the devices. A wireless IoT 'thing' sleeps, senses, connects, and sleeps. The efficiency with which it performs these tasks impact energy consumption, which also involves design trade-off between functionality, size, and battery lifetime (Alan Hendrickson, 2017). Also, many new technologies offer a solution to this problem. The various data transfer protocols for IoT are listed in Table 9.1. The most popular and widely adopted are the Low Energy Blue tooth (BLE), ZigBee, and Wi-Fi and the selection depends on the particular application.

The extended arm of IoT is the AI. The data from the product or system can be analyzed, formatted, detect parametric variations, and further used for a predictive

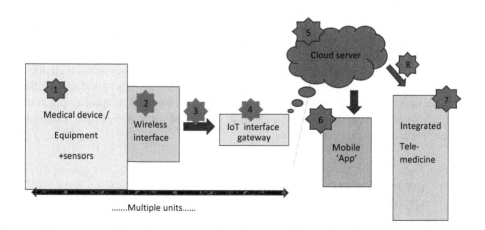

Figure 9.1   A simple block diagram of a simple IoT system in health care.

Table 9.1

Popular Communication Protocols in Data Transfer

Sl no.	Name	Description
1	BLE	Low power Bluetooth, used particularly on wearables
2	IEEE 802.15.4	A low-level wireless personal area networking standard, managed by IEEE
3	ZigBee	A popular IoT networking stack built on top of the 802.15.4 MAC. Used by many home networking devices.
4	Wi-Fi	Wireless protocol widely used on mobile platforms
5	6LowPAN	For IP traffic over lightweight wireless networks
6	LPWAN	Low Power Wide Area Networks. A collection of narrowband cellular technologies that archive wide-area coverage with IoT devices
7	CoAP	Constrained application protocol is an IETF standard design application layer for machine-to-machine communication
8	WoT	Web of Things is similar to CoAP but based entirely on World Wide Web communications.
9	MqTT	Message queues telemetry transport is an ISO standard lightweight, publish-subscribe network protocol that transports messages between devices. The protocol usually runs over TCP/IP

trend using AI. Machine learning concepts together with special algorithms go a long way to support the AI framework in future IoT-based eHealth systems (Rashid Ali et al., 2019). The predictive capabilities of AI are going through a rapid upward trend. Examples of AI in similar situations, but outside human health care, are in agriculture where AI is used to predict the disease in a particular plant when the photograph of the defective leaf or fruit is presented to the AI algorithm. The AI not only predicts the disease but also provides a solution to eradicate the disease. Though the same logic is not applicable to human health, days are not very far, where AI can effectively be used in detecting and treating human diseases.

The other design challenges in the IoT health care system are the understanding of the requirements of medical fraternity and adopting the equipment toward ergonomic and useful outputs. This requires several rounds of discussions with doctors and medical users and after development making clinical trials to understand the requirements. Further challenges are in upgrading old medical equipment which is already time tested in clinical applications to the IoT platform. One of the solutions is outlined in the subsequent section.

9.3   IoT INTEGRATION OF PREVIOUS GENERATION HEALTH-CARE EQUIPMENT

Established industries are adding industrial IoT products and services to their legacy products and their customers are deploying them in environments with a mix of old and new equipment. This calls for a seamless transition to the existing field personnel (Joy Weiss & Ross Yu, 2015).

Previous generation equipment, including medical devices, were mostly analog with few digital outputs and hence it is difficult to implement IoT technology. Most of the rural hospitals and clinics have yet to get modern equipment with digital outputs and, obviously, due to cost implications. Tapping the signals from this equipment will normally have an adverse effect on their performances and problems on warranty.

Hence, another alternative is to scan the screens with a medium-resolution camera and use OCR technology to convert the displays into digital readouts and then use an add-on module like Wi-Fi/Bluetooth or similar technologies to port out the digital data. These data can be stored, analyzed, and structured even for predictive calculations. An overview of the system is shown in Figure 9.2.

The camera is placed in a non-obstructive position in front of the existing patient monitoring system and periodically captures the frames of the digitally displayed values. The processing unit temporarily stores the image and passes it onto the Google Cloud through the Wi-Fi module and a secure Internet connection. The Vision API's OCR method processes and extracts the digital values of the input image. The OCR extraction process from a display of an old patient monitoring system is indicated in Figure 9.2.

In the OCR extraction process, the cropped region of interest (ROI) image is fed into the open source Google OCR project called Tesseract through the wrapper python library-Pytesseract. The Pytesseract method to be invoked is image-to-string () with parameters image_ROI & Config. It is very important to set the configuration of Tesseract attributes such as Language, OCR Engine Mode, and Page Segmentation

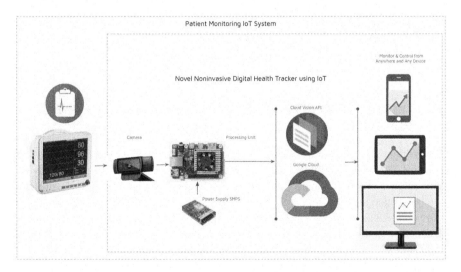

Figure 9.2 An overview of data capture and processing from old health-care instruments using IoT.

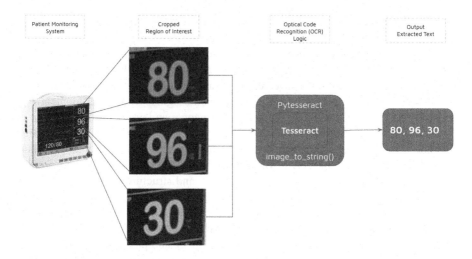

Figure 9.3 Extraction of values from monitor display.

Mode with the best suiting values depending on the patient monitoring system that is being monitored non-invasively. The output from the Tesseract module is the extracted text, that is, in our case, numerical values representing the vitals of the patient being monitored (Figure 9.3).

The extracted digital values are mapped with the corresponding fields such as heart rate, RESP, and so on, and stored in the Firestore database. This remote patient data is then displayed in real time using a web or mobile app for the hospital in-charge to take necessary action. The advantages of such systems are the compatibility with any existing patient monitoring systems or medical devices with display. Capable of extracting data from any digital font size or font color or font, this is a non-invasive technique, easy to implement, and cost-effective.

9.4 DIGITAL TWIN CONCEPTS ON HEALTH CARE USING IoT PLATFORM

Digital twin concepts date back to 2012, intended mainly to assess the product life cycle as well as reliability predictions in aeronautical engineering, specifically for aero-engines. With the same principle applied across, a digital twin in health care can be defined as the digital equivalent of the organ or part of the body by streaming its real term values of the various parameters of its functioning to a separate entity. Health care in digital twin involves breaking the human health data in physical plane and transferring the data to a virtual platform and reassembling the data to replicate the human aspect through IoT concepts. This virtual entity can be analyzed without any inhibitions and various simulations can be done to diagnose and the information can be back transferred to the physical plane to treat the patients.

The digital twin concept in health care provides an environment for diagnosis, therapy, and enhancement. However, it has the potential to impact one's personal identity, raising novel ethical, legal, and social issues (Koen Bruynseels et al., 2018). Hence, it is necessary to have watertight security on personal data, and more precautions are required to safeguard the privacy data of individuals, even in a wireless/digital environment. There are many methods to secure data like multiple levels of encryption during transmission and storage. The research into the implementation of digital twin involves leveraging the latest technologies such as 5G, IoT, Wireless, RFID, Ethernet, Sensors & Actuators, and the Cloud (Jones et al., 2020) (Figure 9.4).

IoT devices' capability of streaming out parametric data in real time is a perfect fit to conceive the digital twin. For example, if you stream the live data of the patient on a ventilator such as his respiratory rate, pressure, volume, and flow characteristics, then we can reconstruct a digital twin of his lungs. The digital twin of the lungs can then be analyzed to improve its functioning using various predictive algorithms and thus be able to diagnose and treat patients easily without doing extensive tests on intricate parts like lungs. Advanced simulation and prediction algorithms can be used to predict future health issues of that particular human using the twin and take precautionary measures. A digital twin will be a boon to the medical fraternity to monitor and analyze the health of the patients remotely, thus saving physical interaction time and complexity of physical interventions. This, in turn, brings down the cost, unnecessary hospital visits and improves resource allocation and planning.

The heart being a blood-pumping organ and an intricate and often needs much care due to our current sedentary lifestyles can be digitally twinned through IoT technology. The ECG and other parameters can be continuously fed to the twin system so that there is a digital twin of the heart which can be analyzed remotely by the medical fraternity to analyze, diagnose and provide pre-emptive treatments. Similarly, most of the intricate organs in our body can be digitally twinned with the parameters obtained through IoT technology and a proactive medical treatment can be made to

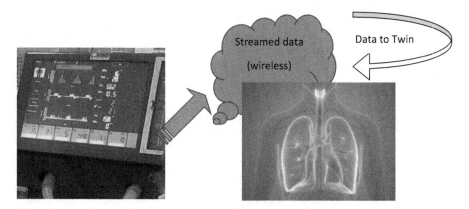

Figure 9.4 Patient connected to ventilator.... streams data...simulates a digital twin of the lungs.

ward off futuristic ailments. However, the challenge for the designers here will be the cross-functional knowledge requirement for both engineering, software and medical functions.

9.5   IoT AND MOBILE APPLICATIONS IN HEALTH CARE

With the advent of 4G/5G and smartphones, the applications in mobile for health-care domain gained importance and prominence. Initially, the non-critical, non-invasive parameters were introduced in the mobile applications like pulse rate, temperature, and so on. The health-conscious generation demanded more parameters like number of steps, calories burnt, nutritional values, and BP, which saw a number of applications in mobile in the health care domain. Some of the advanced mobile health-care applications include cancer symptoms tracking, blood glucose monitoring, smart Insulin delivery for diabetics, connected inhalers, ingestible sensors, wearable asthma monitor, and so on. The IoT platform on these devices helps to take the data to a central place/server/cloud where it can be stored, retrieved, and analyzed from a remote place.

The design challenge in developing IoT mobile applications in health care are the knowledge requirement on medical terms and the software knowledge. Though the medical fraternity can give inputs, the designer should be capable of understanding and depict it correctly in his product or application. Figure 9.5 shows a typical

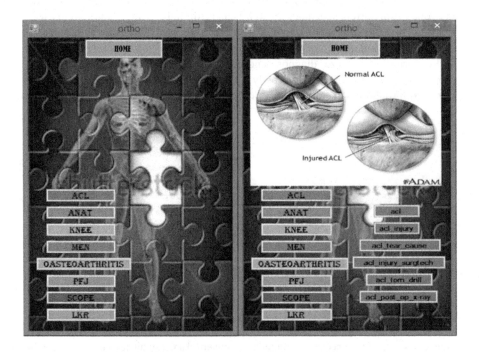

Figure 9.5   A mobile IoT platform development for health care – Orthopedic example.

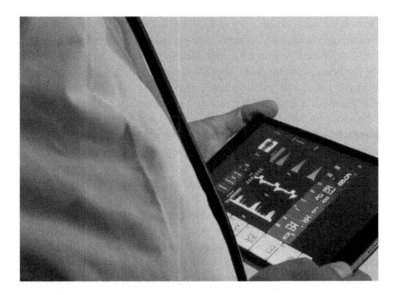

Figure 9.6 Patient parameters viewed remotely by a doctor in a tab.

IoT-based health-care mobile application development for orthopedic knowledge for patients suffering from different ortho problems. It calls for proper knowledge of various problems in the ortho domain and the associated symptoms, and so on for a developer along with his software knowledge.

In critical care units, the mobile application provides a support mechanism for medical attendants to view the parameters remotely, get alarms whenever patients' monitored parameters go critical. It also saves time personally going around the units, especially during night shifts. Some of the applications are designed to view multiple patients on the same page of a tablet and color coded to give the priority of attention based on the criticality. Figure 9.6 provides a view of a medical practitioner looking at a patient's medical parameters from a remote place, which is feasible through IoT technology.

Recently, there is also news on ventilators which run from a mobile application Deepak Agarwal (2019) but need extensive clinical trials to ensure safety for the patients. The advantage of driving a ventilator through mobile is that the data is already in a communication domain and can easily be ported out for IoT applications.

9.6 IoT AND ARTIFICIAL INTELLIGENCE APPLICATION IN HEALTH CARE

The basic data collection on the patient parameters from the medical devices is only a small portion of IoT. Using large volumes of data from complex medical instruments like X-Ray, MRI, CT, PET scans are treasures of information on health issues of people. It is often difficult to analyze the data from complex medical devices with simple technologies, wherein AI steps in to provide a solution.

One industry where IoT and AI, individually or together, are making significant impacts is the health-care industry, which is constantly under pressure to reduce costs while addressing a rapidly growing unhealthy population (Rushabh Shah and Alina Chircu, 2018). With many new diseases and a shortage of health-care professionals, it is all the more useful to have a technology which can continuously monitor the health parameters through IoT devices and have an AI support to predict and diagnose an ailment very quickly and at an early stage. A large amount of data collection supported by AI can predict community spread of diseases and gives an early warning toward preparedness of the government and health-care officials to tackle any pandemic situation before it goes out of hand.

AI not only helps in diagnosis but also predicts a future health concern using the available data which is a boon for doctors and patients. A proactive treatment under such circumstances can save precious lives. For example, an abnormality or malignancy in the lung or other organs can be tracked with different parameters of health conditions using IoT to predict cancer before it is too late to detect and treat because cancers in stages 1 and 2 are easier to treat than cancers in stages 3 and 4.

Patient engagement and adherence have long been seen as the 'last mile' problem of health care – the final barrier between ineffective and good health outcomes and more patients proactively participate in their own well-being and care, the better the outcomes – utilization, financial outcomes, and member experience (Thomas Davenport and Ravi Kalakota, 2019). These factors are being addressed by Big data/AI and supported by IoT devices that track the parameters of the patients after they are discharged from hospitals. The AI can be successful in these cases only if they get continuous data on the health parameters, and this is possible only through IoT devices and platforms.

9.7 SENSORS FOR IoT APPLICATION IN HEALTH CARE

The critical component of an IoT system is the sensor. Sensors are devices that detect physical, chemical, and biological changes in signals and provide a way for those signals to be measured. It is the first physical layer that converts the physical parameter into an electrical signal, be it voltage or current or resistance variations.

In health-care IoT devices, sensors can monitor temperatures, pressures, pulses from heartbeat, electrical, chemical, and biological levels of users and/or patients. An overall block diagram of the IoT in health-care system is depicted in Figure 9.7 with the highlights of sensors, which are critical for any system. In general, the output sensors are variations in current, resistance, and micro voltages. This needs to be amplified and converted suitably to voltage variations of higher magnitude. Then comes the Analog to Digital conversion (A to D), which converts the analog voltages to digital signals for the successive sections to handle. The A to D resolution is quite important for the quality of signals. The resolution is typically expressed in bits viz., 8 bit, 10 bit, 12 bit, and so on. Though a better resolution is more accurate, the higher the resolution, the higher the size of the digital signal and calls for higher

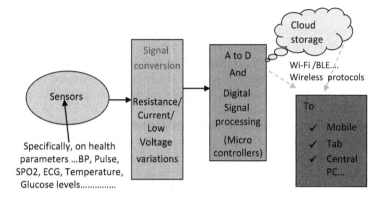

Figure 9.7 Overall view of a typical health-care IoT system with typical Sensor outputs.

bandwidth of the transmission sections. Hence, normally, there is a trade-off between the resolution and the transmission capability.

The challenge of sensors in the health-care IoT system is the need for a minimally invasive device that is easy to connect, run, and maintain. Also, the materials used in sensors which are in contact with the user/patient should not have materials which are allergic. Yet another factor for the sensors is the low power consumption requirement since they will be used continuously and the design should take care of the long-term reliability of the sensing element. As the health-care market is cost-sensitive, it is necessary to make these sensors cost-effective.

There is still a lot of scope and demand for sensors which can measure the health parameters of the humans. MEMS and nanotechnology seem to be the futuristic domains for these sensor designs since they are relatively small, sensitive, and can be made selective to parameters. Flexible electronics is another emerging domain, where the circuits are printed or formed on flexible polymer sheets or even textile materials which act as sensing/signal-processing units which can be worn around the body or part of a user. As the technology of electronic packaging industry is getting stronger, some designs integrate signal processing and communication as a part of the sensing element. The power for the devices is even derived using printed piezoelectric materials, which can generate small power during the motion of the individual since the power requirement is very small for these devices. Even the wireless transmission requirement for the IoT part is handled by the printed antennas which are part of the flexible electronic units. The latest technologies in sensors like voice-activated modes, gesture activation modes, and so on, and make it much easier for users than pushing buttons and selecting settings.

9.8 IoT AND ITS SECURITY ASPECTS

Like all technologies, IoT has both positive and negative aspects. Especially coming to the health-care sector, the sensitivity of data and privacy details is high and needs to take utmost precaution and measures to have it secured. In a health-care environment, the breaches in security not only affect the privacy, but sometimes may be risky for the patients as well. Imagine a scenario where a hacker swaps the records of a healthy one with that of a sick patient; and that will play havoc in both their lives.

The security solutions to detect, prevent and respond to intended or unintended data breaches should cover hardware, software and the connectivity. The health-care IoT devices/systems should use hardware, which has inbuilt security protections like symmetric cryptography and the software should have multiple layers of authentication, anti-cloning algorithms, and the connectivity should be tamper proof. There are also microcontrollers with built-in Internet connectivity and software layers to provide a secure wireless transmission. Also, the device manufacturers avoid default passwords to start the unit like '0000' which normally is not changed by the user for a long time, inviting cyberattacks on the devices. The latest devices are designed for self-software updation, and any security loopholes identified later in software are automatically patched up online. The use of hardware locks, secure booth mechanisms and control of privacy data with users are some of the measures to ensure security in health-care IoT devices. Typically, a multiple set of security measures are incorporated into the IoT design in multi-layer architecture. The general measures of security in health-care IoT systems include multiple layers of strong encryption, access control, integrity checks, firewalls, and self-reporting in case of breaches.

The software reliability of the system can be tested using reliability tools and that can ensure a certain amount of data security considering the standards followed in writing the software codes. The major constituents of the security aspect of the systems are confidentiality, integrity, and Authenticity. There are emerging new technologies for security protection in the data handling domain and the designer needs to adopt and apply these new concepts to make the device safe against cyberattacks and data breaches.

9.9 REMOTE MONITORING USING IoT IN HEALTH CARE IN THE CONTEXT OF COVID-19

We live in a world where technologies tend to make our lives easy. Health care is a domain that needs the support of the latest technologies to make diagnosis, monitoring and treatment easy and affordable. There has always been a shortage of health-care professionals and especially when there is a pandemic situation. New technologies like IoT and AI provide a lot of support in those situations to the medical fraternity. Before the IoT phase, the patients had to personally visit the hospitals or clinics and also there was limited continuous monitoring of patients. IoT in health care is gaining importance due to its obvious benefits, by streaming data out of the medical devices to a separate entity for further analysis, diagnosis and treatment. Further, the IoT technology in health-care devices supports pandemic environments (like the COVID-19) wherein the patient parameters need to be monitored from a remote place for avoiding infections of the medical fraternity.

IoT supports preventive medical care as well by data accumulation from daily checks on self-care medical devices such as BP monitors, glucose monitors, and so on, which can analyze the day-to-day variations and alert the medical support system on the variations and alarming trends. A camera-based IoT system at home for the elderly can be a good option to monitor their activities, remind them of medication requirements, and provide support for their easy living. Also, there is cost-effectiveness in using IoT in health care by an increase in preventive care compared to emergency care and also the hospitalization time can be minimized with the support from hospital staff after an early discharge using streaming data from the patients with connected specific IoT devices. The IoT in health care further reduces the burden on insurance companies by providing easy validation of claims and reward people who are able to take precautionary health measures using IoT devices. The collection of real-time data also provides a lot of support for further research and development of new treatment methods. However, while personalized health care has huge benefits, the IoT technologies still face many challenges due to the need for cost-effective sensing technologies, advanced algorithms, life-logging data, methods of coping with uncontrolled environment, and high volume of data set, security and privacy, and so on (Jun Qi et al., 2017). This in turn provides a large scope and competitive environment for design and computing engineers to overcome these challenges.

9.10 CONCLUSION

The health-care IoT covers a variety of activities like remote monitoring of patients' critical parameters, tracking of drugs intake by patients, prediction and proactive treatment, ambulance telemetry, and hospital asset management including patient tracking after discharge. Further, the health-care IoT technology enables the latest generations of hospital equipment to transmit information directly to the electronic health records of patients from home devices. Connected sensors and machine-to-machine (M2M) communications can report on dosing, equipment condition, and present patient health. Significantly, these happen in real time, detecting errors and breakdowns and health crisis events, even sending alerts through SMS messages to health-care professionals when there is a critical issue with the patient. The major drivers for IoT in health care will be the increased health consciousness of individuals, growing popularity of wearables, higher demand for remote health monitoring like in rural areas, and the new business model in marketing the new technology products in health care. However, the success of the health-care IoT lies with the adaptation and acceptance of the new technology by medical fraternity as there seem to be a lot of pressure from the environment to adopt in full or at least in part.

ACKNOWLEDGMENT

The author is grateful to Prof. P. Radhakrishnan, Director, PSG Institute of Advanced Studies (PSGIAS) for the support and encouragement. The author thanks Manivasagan Ramamoorthy, Chief Data Officer, Vyazhan Technologies Private Limited, Coimbatore, for his valuable inputs for this chapter.

REFERENCES

Agarwal, D. (2019) Retrieved from: https://fit.thequint.com/fit/aiims-worlds-cheapest-portable-ventilator, Feb 2019.

Ali, R., Qadri, Y. A., Zikria, Y. B., Umer, T., Kim, B. S., & Kim, S. W. (2019). Q-learning-enabled channel access in next-generation dense wireless networks for IoT-based eHealth systems. *EURASIP Journal on Wireless Communications and Networking, 2019*(1), 1–12.

Bruynseels, K., Santoni de Sio, F., & van den Hoven, J. (2018). Digital twins in health care: ethical implications of an emerging engineering paradigm. *Frontiers in Genetics, 9*, 31.

Davenport, T., & Kalakota, R. (2019). The potential of artificial intelligence in health care, *Future Healthcare Journal 6*(2), 94–98.

Guo, Y., Hu, X., Hu, B., Cheng, J., Zhou, M., & Kwok, R. Y. (2017). Mobile cyber physical systems: Current challenges and future networking applications. *IEEE Access, 6*, 12360–12368.

Guth, J., Breitenbücher, U., Falkenthal, M., Fremantle, P., Kopp, O., Leymann, F., & Reinfurt, L. (2018). A detailed analysis of IoT platform architectures: Concepts, similarities, and differences. In *Internet of Everything* (pp. 81–101). Springer, Singapore.

Jones, D., Snider, C., Nassehi, A., Yon, J., & Hicks, B. (2020). Characterising the digital twin: A systematic literature review. *CIRP Journal of Manufacturing Science and Technology, 29*, 36–52.

Kim, K. D., & Kumar, P. R. (2013). An overview and some challenges in cyber-physical systems. *Journal of the Indian Institute of Science, 93*(3), 341–352.

Shah, R., & Chircu, A. (2018). IoT and AI in healthcare: A systematic literature review. *Issues in Information Systems, 19*(3), 33–41.

Truong, H. L., & Dustdar, S. (2015). Principles for engineering IoT cloud systems. *IEEE Cloud Computing, 2*(2), 68–76.

Weiss, J., & Yu, R. (2015). "Wireless Sensor Networking for the Industrial IoT", Electronic Design, Sep 2015.

Yang, P., Qi, J., Min, G., & Xu, L. (2017). Advanced internet of things for personalised healthcare system: A survey. *Pervasive and Mobile Computing, 41*, 132–149.

10 Applications of IoT in Health Care: Challenges and Benefits

Kedar Nath Sahu
Stanley College of Engineering and Technology for Women
Hyderabad, India

Ravindharan Ethiraj
Osmania University
Hyderabad, India

Paramananda Jena
DRDO
Bengaluru, India

CONTENTS

10.1 Introduction 177
10.2 IoT in Health care 178
    10.2.1 Need for IoT in Health care 179
    10.2.2 Importance of IoT in Health care... 179
    10.2.3 Benefits of IoT Based Health care 180
    10.2.4 Challenges of IoT-Health care 181
10.3 Existing Devices Used in IoT-Health care.. 182
10.4 The Internet of Nano Things 183
10.5 Case Studies in IoT Based Health care... 186
    10.5.1 IoT-based Health care in India... 190
10.6 Conclusion.. 191
References... 192

10.1 INTRODUCTION

One of the biggest technological outbreaks that marks the beginning of the era of the modern-day industry which aims at enhancing the Quality of Service (QoS) is the Internet of Things (IoT), a cooperative or interactive wireless communication technology. An improved QoS, as depicted in Figure 10.1, primarily desires to enable access to and retrieval from the physical and digital world of information about the

177

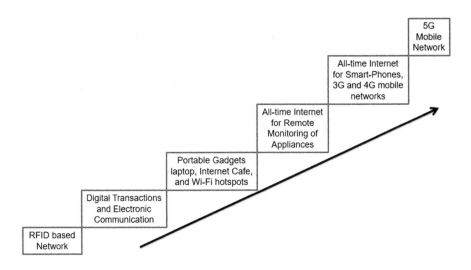

Figure 10.1 Overview of technological growth toward the growth of QoS.

physical objects and persons tagged on to a radio-frequency identification (RFID)-based network. This has led to an emerging need for enhanced capabilities for digital transactions and electronic communication. The demand for "portability" was the next desire for comfort. Soon, laptop as a pocket calculator, Internet cafe, and Wireless Fidelity (Wi-Fi) hotspots became a reality. With miniaturization continuously happening, the next vision was to have the Internet at all times over the palms of our hands. This came true when smartphones as well as third generation (3G) and fourth generation (4G) mobile networks were launched into the mainstream of people's lifestyles. Then, the focus shifted to home appliances, security systems, cars, machines, factories, and workplaces; soon, other infrastructure got connected to the Internet. With the fifth generation (5G) of mobile broadband communication technology on the horizon, very soon everything could possibly be connected.

This chapter is organized into five sections apart from Section 10.1. Section 10.2 reports the need, importance, benefits, and challenges of IoT-based health-care system. Sections 10.3 and 10.4 present the existing devices for IoT in health care and the Internet of Nano Things (IoNT), respectively. Section 10.5 presents the case studies on three recent research projects aimed at promotion of sustainable work for EU industries. The summary is discussed in Section 10.6.

10.2 IoT IN HEALTH CARE

There are increasing consciousness and involvement of people all over the globe with regard to their health. The much needed remote as well as home care possibilities with reduced cost and fast medical services result in a large number of players in the health-care ecosphere emerging with novel propositions and collaborations. Globally, more than 60% of health-care establishments have adopted the IoT devices into their resources, according to the web article "Internet of Things (IoT) in health care: benefits, use cases and evolutions" (https://www.i-scoop.eu/internet-of-things-guide/internet-things-health care). Certain regions, for example, the United States, are the prime users of IoT in health care and leverage its key IoT devices for health data. The administration of IoT gadgets in health-care services is alternately referred to as the "Internet of Healthcare Things (IoHT)" or "Internet of Medical Things (IoMT)".

10.2.1   NEED FOR IoT IN HEALTH CARE

Thousands of people all around the world lose lives due to being deprived of quality health care, especially in rural areas, as well as lack of timely treatment and concern about the likely dangerous impact of a disease. According to the 2003–2008 report of UNICEF, the maternal mortality ratio that refers to the death of a pregnant woman is 2.5%. According to the World Health Organization (WHO) and UNICEF, there are 5,85,000 deaths related to pregnancy and childbirth (Rohokale et al., 2011). In addition, the world has witnessed a rapid surge in bulging population, in general, and aging population, in particular, which regularly adds to the commonly existing burdens of (i) insufficient infrastructure, (ii) fewer health-care experts, (iii) inferior quality of being able to reach standard medical facilities in the rural sector, (iv) costlier treatment, and so on. These pose serious impediments to progress in health care, thereby making the health-care system itself more and more ailing. Thus, there is a growing need for a more efficient health-care system. This can be achieved by considering all entities involved in the health-care system connected over a common network. This gives rise to the applied field related to the *Internet of Healthcare Things (IoHT)*, a technology-based on a self-configured network that provides connectivity by integrating every element of a health-care system.

10.2.2   IMPORTANCE OF IoT IN HEALTH CARE

In general, any IoT-enabled health service refers to the integration of *things* involved in medical activities such as smart devices, sensors, relevant health-care apps, artificial intelligence (AI), and so on, using the Internet. It is fundamentally expected to offer a digitally caring environment for all of us by maintaining the following attributes:

- It must keep the essential framework of medical apparatus, application software like the operating system, health-care utilities consistently connected.
- The application must be user-friendly and caring by sending message alerts, reminders, personalized tips, advisories, and recommendations regarding suitable diets, do's and don'ts.
- It must be able to perform continuous monitoring of the patient's health condition and provide varying treatment suggestions following a thorough test on the perception level of the body to further medication. IoT must be vigilant about the patient's health to provide information with regard

to the response and symptom variations of the patient. In this connection, electronic health records (EHRs) are largely essential.

Instantaneous observation through the associated link of devices plays a major role in saving human lives during the medical emergencies such as cardiac arrest, bronchial asthma, chronic diabetes, and so on. The connected devices can be smart medical devices enabled with the smartphone apps, which fetch medical and other useful health data and apply the smartphone data link to convey the information to the medical practitioner. For example, any smart insulin pen records the time/amount/type of insulin in a given dose. The pen not only helps calculate the dosage but also keeps track of the injected data. When paired with the smartphone app through Bluetooth, a pen delivery system keeps track of the insulin intake history and other helpful information, which in turn gets communicated to the physician automatically. Another fascinating application of IoT is its ability to control medication intake and detect the patient's response to the use of a particular medicine. The patient operates a device by taking prescribed medication using an ingestible sensor. Consequent to the arrival of the sensor at the stomach, a signal is transmitted to the patch worn by the patient. As depicted in Figure 10.2, the digital information is directed to the mobile phone of the patient, followed by its transfer to the *Cloud* at which the health advisors do the needful.

10.2.3   BENEFITS OF IoT BASED HEALTH CARE

The most important beneficiaries of IoT-enabled health care include doctors, hospitals, health insurance companies, and pharmaceutical companies that prepare necessary medications compliant with the ongoing treatments. Health-care systems using IoT help optimize the medication administration activities, undertake medical prescriptions with precision, assist in monitoring of patients from a distance, medical as

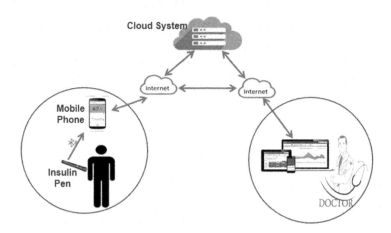

Figure 10.2   Example depicting importance of IoT in health care.

well as paramedical staff, and so on. Some of the crucial benefits that can be derived from IoT-enabled health services are outlined below.

    i. **Optimized caring process and enhanced quality:** Regular tracking of the medical staff, right medication at the right time, resources urgently needed, hospital beds, and so on are highly cumbersome and challenging in every health-care system, which would otherwise lead to endanger the patients. Due to easy access to health-care data in an IoT framework, it can be possible to achieve concurrent exchange of true real-time data and monitoring of health parameters during the treatment of a patient. As a result, the treatment period can be shorter, and hence, coverage of patients may be larger. A patient can be kept in vigil by monitoring through sensors, right from their place of stay until they reach the health center. The real-time health data can be automatically updated or added to patient records without the presence of nurses appointed to reading or updating the patient health record.

    ii. **Lesser expenses and errors:** As an IoT-based health-care system can provide medical personnel with accurate health data, this might result in reduced medical errors. Consequently, it will help reduce the number of unnecessary hospital visits, stays, and frequent admissions. This will facilitate the effective utilization of resources and hence, less expensive care.

   iii. **Enhanced rural access:** IoT-based health monitoring can offer rural reach for quality health care so that patients can avail remote consultation of physicians serving in multispecialty hospitals in urban areas.

However, to support a patient-centric approach and far-reaching benefits, IoT-enabled health care must be supported by robust infrastructure with reliable Internet facility, less costly IoT devices, data privacy and security, and so on.

10.2.4   CHALLENGES OF IoT-HEALTH CARE

At every stage of an active system, there are likely issues and challenges of concern. These include data storage, handling, and safeguarding the health data to protect patient's privacy, data privacy during its exchange among various integrated devices for time-to-time processing, protection for data transparency according to data security guidelines for data usage and accessibility. Effective data management solutions with regard to recording, sharing, and processing of data are inevitable to predict and avert health complications, use clinical procedures, apply drugs, and provide optimum treatment to the patient.

    Security Issues: However, the IoT-enabled hospital systems are still in experiment, and hence, it is dangerous to take shelter there because they might face serious errors arising out of IoT network management. The incident of Pablo Garcia, a teenager who received 38.5 times the dosage he should have received due to a setting error in the EHR system, is an example with regard to the EHR based on a computer-administered medication management.

    As the usage of IoHT grows, it multiplies vulnerable points through a swarm of endpoints, networks, and channels (The Medical Futurist, 2019). It is reported that medical devices are highly vulnerable to being hacked and are becoming the largest threat to IT-enabled health-care services. Serious security threats are widespread in the case of X-ray systems, blood refrigeration segments, CT scanning apparatus, implantable cardioverter defibrillators (ICD), and implantable neurological devices. In addition, issues such as technical preparedness, regulatory and legal aspects in connection with data exchange beyond jurisdiction must be considered; moreover, the economic viability of implementation in developing countries cast very important challenges to explore the full potential of IoT in health care. At present, continuous development is happening in the areas of new medical devices, smartphone apps, sensor-based equipment, and solutions by various start-ups. In this connection, crucial guidelines for manufacturing medical equipment as well as industry standards are inevitable, failing which, will lead to the greatest risk of use.

10.3 EXISTING DEVICES USED IN IoT-HEALTH CARE

The application of IoT in health care proved real when devices were connected to pick up breathing rates and monitor sleeping patterns, mobility, and gait, in order to provide information about a patient's emotional state from their heartbeats. The existing IoT-enabled devices handling health-care data include modern smartwatches, physical fitness bands, wearable monitoring patches, sensors for detection of the abnormal rhythm of the heart, improved access and medical compliance through creative smart pill gadgets, automatic drug dispensing setups, the combination of telemedicine and IoT (Lu & Liu, 2011), and so on. A smartwatch is able to forward information about vital signs to a physician's tablet. Toilets enabled with microchips similar to the MC10 biostamps are able to monitor urine so that a patient doesn't have to physically carry a urine sample to the doctor. Sensor-based log for movement patterns, tracking of water usage patterns using bathroom sensors, and measurement of basic vital signs using digital mirrors are some more existing developments of IoT-based health-care systems. The Autobed system developed by GE and introduced in Mount Sinai Hospital proved very effective and helped decrease the waiting period of patients in emergency rooms significantly by 50%. This is based on an algorithm that quickly decides the bed on which an incoming patient is to be placed. The algorithm utilizes the status of available beds facilitated by real-time location awareness devices such as the RFID tags, infrared (IR), and computer vision.

As reported in the web article "A Digitally Caring Environment: The Internet of Things in Hospitals" (The Medical Futurist, 2019), NYU Langone's newly opened hospital offers MyWall, which allows the patients to supervise their diurnal needs like placing orders for meals, watching fun, and so on by means of tablets. Patients are able to know about the team appointed for their care and adjust room lighting or temperature by themselves. The IoT device "iN" brought out by a New York-based start-up *Inspiren*, worked with a focus to mitigate the negligence of patients. This device, based on technologies such as computer vision, deep learning, and recognition of natural body movement of humans is able to discern simultaneously the presence of medical staff, assess the risk of environmental safety, collect and aggregate data from other auxiliary medical devices such as electrocardiogram (ECG) or vital monitors, detect temperature, noise, brightness, and so on from environment safety sensors. Thus, the device can aid to monitor the behavior of both medical staff and patients that in turn enhances the healing process. IoT networks based on RFID are used to track the (i) medication and (ii) inventory and maintenance of medical equipment.

The real-time location system (RTLS), a setup that can be considered as a kind of "indoor GPS" for hospitals provides real-time monitoring and supervision of medical personnel and appliances. This also works similar to the RFID technology as the system incorporates location sensors placed on various assets such as medical staff or equipment and the patients. Philips is working on an e-Alert system that can monitor the medical kits and send alert notifications to the hospital workers, if needed.

Several cooperative IoT models applicable for Adhoc wireless and wireless sensor networks have been reported in the literature (Rohokale et al. 2011; Liu et al. 2006; Nosratinia and Hunter 2004; Sadek et al. 2007; Kailas and Ingram 2010). The cooperative mechanism used by Rohokale et al. (2011) in their IoT model for health care is based on continuous monitoring of human health parameters, namely, blood pressure, hemoglobin, blood glucose level, and so on. Furthermore, the model shows to be reliable and energy-efficient.

The "iMed box" with the sensors plugged onto the iMed patch attached to the patient's fingertip is shown to dispense the patients with doctor prescribed medicines (Yang et al., 2014). Washable smart clothing (Chen et al., 2017) consisting of various health sensors and cloud service are designed for use by patients. Smart beds are able to collect health data when a patient is asleep (García-Magariño et al., 2018). A product summary pertinent to the above existing IoT-enabled devices is presented in Table 10.1.

The health-care sector is likely to get a new dimension with the growth in the field of nanotechnology (Akyildiz & Jornet, 2010a) that encompasses synthesis, characterization, and execution of nanomaterials and devices. As reported in the article top five trends in nanotechnology (Giges, 2013), nanotechnology is becoming the most exciting development in the biomedical domain, where growth is happening in both diagnostic and treatment areas, and therefore, nanomedicine is trending as one of the top five in the nanotechnology domain. It is only natural that nanotechnology gets linked to IoT.

10.4   THE INTERNET OF NANO THINGS

The IoT, undoubtedly, is the metamorphosis of the Internet application, communication, as well as exchange of data among devices, sensors, and objects. It has pushed itself to a fascinating domain under the name, IoNT, which essentially connects things based on nanotechnology. The term, "Nanotechnology" was instituted by Norio Taniguchi in 1974 in course of a scientific convention (Akyildiz & Jornet, 2010a). The proposition of building "small" things is typically attributed to Richard Feynman built on his talk delivered during 1959. He has mentioned that machines would fabricate smaller machines and further products in unison with atom-by-atom control.

Table 10.1

Summary of Products Related to IoT-Enabled Devices

S. No.	Year of Introduction	Device	Reference/Developer	Functionality
1.	2018	MyWall, a 75-inch high resolution electronic display screen	NYU Langone, US	Supervision of diurnal needs.
2.	2017	"iN"	Inspiren, US	Mitigation of negligence of the patients.
3.	2017	Ability MyCite-A creative smart pill	Proteus Digital Health,US & Otsuka Pharmaceutical Co., Ltd., Japan	Tracking of drug regimen compliance from inside.
4.	2016	e-Alert system	Philips Healthcare Innovation Centre (HIC), Pune, India	Monitoring of medical kits, imaging and diagnostic equipment.
5.	2016	Chem–Phys patch	University of California-San Diego, US	Concurrent real-time measurements of health and fitness.
6.	2013	Autobed system, Mount Sinai hospital in New York	GE Healthcare,US	Reduction of waiting period of patients in emergency rooms.
7.	2011	Medical mirror	Ming-Zher Poh,Massachusetts Institute of Technology,US	Vital sign monitor.
8.	2011	Real-time location system (RTLS) at Atlanta Hospital	Piedmont Healthcare, Georgia, US	Real-time monitoring and supervision of medical personnel and appliances.
9.	2010	Sensor-based log	Aquaone Technologies, Westminster, US	Monitoring of water usage patterns.
10.	2010	Toilets enabled with microchips similar to the MC10 biostamps	Toto Ltd., Japan	Monitoring of patient's urine.
11.	1998	Smartwatch-Ruputer	Seiko, Japan	Information transmission about vital signs to physician's tablet.
12.	1984	Bathroom sensors/Sensor-activated faucets	Oliver N. Wareham, Australia	Tracking of water usage.
13.	1977	Wireless ECG heartrate monitor	Polar Electro, Kempele, Finland	Measurement of electrical heart information.
14.	1971	Automatic drug dispensing setup like portable digital tablet counter	John Kirby, Manchester, England	Help pharmacies in counting medications accurately.

There is a little dissimilarity in the particle characteristics in terms of comparing the size of particles in solid matter in visible scale with the size observed by means of a regular optical microscope, according to a web article with respect to Nanoscale reported by the National Nanotechnology Initiative (NNI), a U.S. Government R&D initiative involving nanotechnology. However, a significant variation will be taking place with the material properties, when particles produced are of sizes as low as in the range of 1–100 nm. Corresponding to this size scale, referred to as the *nanoscale*, which can be observed only with potent and effective microscopes, the properties of particles are influenced by the so-called quantum effects. It is this scale range at which the properties of materials such as electrical conductivity, magnetic permeability, fluorescence, melting point, and affinity of chemical reactions vary with the size. It's not that nanotechnology simply works at smaller dimensions; but it is working at the nanoscale range to allow the utilization of the unique naturally occurring properties of materials, be it physical, chemical, mechanical, or even optical that takes place at this scale. Thus, nanotechnology became an emerging technology that aims to control the structure of engineering materials at nanoscale dimension. This will enable the nanoscale components to perform defined tasks, taking advantage of the specialized properties on that scale. Data storing, sensing, computing, and actuation took a totally different dimension. All the nano components will be integrated into a single advanced nanodevice, which can handle complicated tasks following a distributed approach. Nanotechnology has yielded numerous high-grade solutions to real world domains of implementation like Medicine, Biomedical, Biotechnology, Industry and Cultivation (Akyildiz & Jornet, 2010; Akyildiz et al., 2008b). The backbone of nanotechnology is the nanomachine, which is elucidated as the building block integrated through nano components to accomplish elementary tasks like sensing or activating. It is essential that nanomachines have constructive coordination and cooperation among themselves in order that a wide variety of applications can be resulted in terms of complexity and operations. Thus, in brief, it can be outlined that the amalgamation of nanoscale appliances with the conventional high-speed Internet-enabled communication networks have paved the way for a novel expansion (Akyildiz & Jornet, 2010a, b; Balasubramaniam and Kangasharju, 2013)

To get an idea of how properties change at the nano level, let's take a look at nanoscale gold, which exemplifies the noteworthy properties that arise at this dimension. Particles fabricated from nanoscale gold look to be red or purple, unlike the familiar yellow color. The movement of the electrons in the atomic structure of gold is confined at nanoscale. As a result of such a restricted movement of electrons, reactions of gold nanoparticles with light are different as compared to those of large-scale gold particles. The exclusive features such as size and optical properties have led to important hands-on applications of nanoscale gold particles that they selectively pile up in tumors and permit both precise imaging and targeted laser destruction of the tumor without harming the healthy cells. This is indeed a highly beneficial tool in the hands of health-care professionals and when linked to IoNT would become a boon to patients.

The architecture of IoNT meant for health-care applications (Omanović-Mikličanin et al., 2015) essentially includes the components such as (i) nanomaterial-based nano-nodes, (ii) nano-routers, (iii) nano-micro interface devices, and (iv) gateway for enabling remote control of the system. IoNT-enabled health-care systems need massive data handling that is likely to emerge from nano-networks. The development of transmission and interface gadgets compatible with linking the nano- and micro-scale networks and the latest service models pertinent to application layers for data handling will be the frontline challenges. Incompatibility and non-interoperability of networked hardware, software, firmware, and extinct technologies are common challenges. With technological updates from time-to-time, IoT and, in particular, IoNT, are gaining momentum to provide holistic, preventive, and curative solutions for human health using nanodevices (Freitas, 2005) and nanorobot (Swan, 2014). It must be possible to get quick information about multiple vital health parameters such as body weight, body temperature, heartbeat rate, respiratory rate, blood pressure, blood sugar, and so on in one single and specific application, and IoNT promises just that!

10.5 CASE STUDIES IN IoT BASED HEALTH CARE

Three distinct case studies in support of the research endeavor aimed at promoting the notion of "sustainable work" for EU industries (Pateraki et al., 2020) are discussed here. Each of them refers to the recent research projects such as (i) SMART BEAR, (ii) sustAGE, and (iii) XVlepsis and proposes different investigation perspectives of IoT in health care.

i. **The SMART BEAR Consortium:** The SMART BEAR project (Pateraki et al., 2020) is an intelligent and personalized digital solution for sustaining and extending healthy and independent living. It is planned to be affordable for larger section of the society. In addition, SMART BEAR is secured to operate in a dense network of IoT-based devices, biosensors, and health network. The design of SMART BEAR is meant for independent living of elderly people suffering from hearing loss, cardiovascular diseases, cognitive impairments, mental health issues, balance disorder, and frailty. These objectives can be achieved by evidence-based interventions on the lifestyle of the elderly people. The necessary interventions include the continuous monitoring of medical environment, integration of sensors and assistive devices, and analysis of data of the activities of elderly people so as to provide personalized health care for independent living.

It connects hospitals, dispensaries, and medical services to obtain specific data to make decisions for interventions, as depicted in Figure 10.3. Big data analytics and learning capabilities are utilized for huge data processing in order to produce evidence required for making personalized interventions. Testing and validation are planned through five large-scale pilots involving about 5,000 people living at home

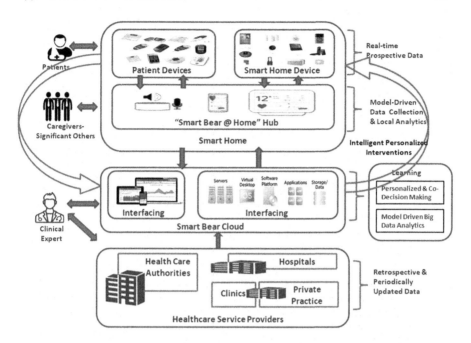

Figure 10.3 Overview of SMART BEAR platform. (Pateraki et al., 2020.)

in Greece, France, Spain, and Romania. These pilots are front runners for the evaluation of the platform to provide health-care services across various regional, state, and EU levels.

 ii. **sustAGE –A smart solution for sustainable work:** This is an IoT healthcare solution (Pateraki et al., 2020) meant for the workers engaged in factories and ports in the European countries. It addresses health problems in three different levels such as (i) improvement of occupational safety and health through risk assessment and prevention strategies based on workplace- and person-centered health surveillance monitoring, (ii) promotion of recommendations for physical and mental health improvements, and (iii) decision-making support related to task or job and role modifications for optimized productivity.

 a. **Application in the manufacturing industry:** The manufacturing sector experiences a yearly increase in the number of elderly working people aged more than 55. Despite automation, it is not possible to perform all works using the machines alone, and hence, the end assembly line work is still being carried out manually. Physical abilities are essential to perform final assembly operation, and cognitive skills are required to cope up with organization strategies and job rotation. A large number of tasks in manufacturing industry differ in posture, workload, and complexity needs both in terms of physical and cognitive strengths. In the automotive industry, different models and small toleration of errors in customized products have been produced. To choose the best match between task and worker in repetitive short-cycle as well as complex tasks, assessment of physical abilities and the mental skills of the workers are essential. It is also important to monitor the environmental conditions and health of the workers in order to derive information on the individual workload that might affect their physical and mental state. Further, actions such as repetitive movements, bend and twisted postures, pushing/pulling/lifting an object along with temporal aspects of the actions need close observation.

b. **Transportation and logistics:** The routine activities of a seaport include (i) loading, unloading, transport, and storage of goods; (ii) movement of the containers of liquids, solids, fractionated products, and so on; (iii) roll-on/roll-off as well as pilotage; (iv) workboat and tug operation; (v) ship repairs; and (vi) vessel traffic management. The workers are usually exposed to stressful and dangerous working conditions. Further, the shift duties affect their sleeping hours leading to misalignment of circadian rhythms, drowsiness, and so on that causes performance deficit. The workers do their work in dusty, less light, and sometimes in flood situations. As they use more physical strength for doing the handwork, they are more prone to musculoskeletal disorders. The demand for high physical and mental concentration at work beyond the capacity of an individual results in negative performances. The architecture of sustAGE system meant for industrial workers is reported in the web article "Integration of the sustAGE system" (https://www.sustage.eu/?p=1570).

The sustAGE monitors the container crane operator, workers involved in the loading/unloading procedures as well as other moving objects/humans in proximity to the crane during maneuvering (Pateraki et al., 2020). It uses environmental sensors, cameras, beacons, GPS, wristwatch, and smartphone devices for generating as well as monitoring data related to the micro movements of the workers. The IoT platform (i) monitors the behavioral actions in the work environment and personal life, (ii) analyzes the users' activities, and (iii) memorizes important episodes. These are aimed at obtaining vital information related to (i) past human activities and states, (ii) aggregation of past user-specific knowledge comprising of user preferences, (iii) results of user performance in work- and training activities, and (iv) long-term abstractions that will allow a more complete physical, mental, and psychosocial user profiling. Recommendations with respect to three different levels, namely, physical, mental, and workforce are provided for continuous monitoring.

iii. **XVlepsis-An intelligent noninvasive bio-signal recording system for infants:** This is aimed at the prediction of medical complications of infants (Pateraki et al., 2020). Detection of medical emergencies of an infant during sleep using unobtrusive and noninvasive systems has a huge medical significance. The use of invasive devices and sensors might even disrupt the sleep of the infants. In the recent past, there are efforts to improve the patient monitoring for facilitating error-free clinical decisions in order to enhance health care quality of infants. XVlepsis IoT platform is particularly useful in the intensive care unit (ICU) of the hospitals. A great deal of effort is required to provide multimodal real-time neonate monitoring platforms, which are ubiquitous and non-obstructive while being at home. As shown in Figure 10.4, it uses a scalable system that includes a smart bed mattress with a camera positioned to monitor the infant cradle without disturbing the infant's sleep. It can also detect possible pathological conditions. The smart sensors are used for the following purposes:

- Recording of video with a high-resolution camera and audio by using a high-definition microphone.
- Ballistocardiogram (BCG) recording (Giovangrandi et al., 2011), which uses pressure sensors under the bed mat to monitor sudden blood ejections into the great vessels for each heartbeat. This generates a plot that represents repetitive body movements during sleep.
- Temperature and humidity detection can be done using suitable sensors.

xVLEPSIS can detect potentially hazardous pathological conditions, with the use of sophisticated machine learning techniques which will lead to a *mobile- or*

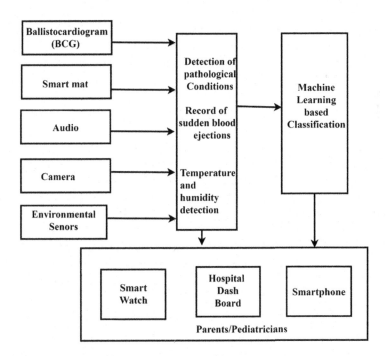

Figure 10.4 Overview of XVlepsis, a noninvasive system for infants.

smartwatch-based notification system to alert the parents in case of emergency and the continuous bio-signal recording, throughout the infant's sleep. The recorded data could be sent to the doctor or the hospital, in case an abnormality is detected, or they could be evaluated by the doctor during regular infant examination, in case nothing critical is detected. Therefore, the pediatrician will be able to examine and evaluate all the available medical data and detect any incident that may have occurred at night without the perception of the parents. The salient features of the development of xVLEPSIS platform are mentioned below:

- Continuous recording of high definition video and audio shall yield more effective monitoring of infants.
- A non-invasive monitoring system shall aid the diagnosis and proper treatment of medical disorders such as, febrile convulsions, epileptic seizures, or apnea that might occur in the absence of the infant's parent.
- As the pediatricians usually face a challenge like evaluating the medical incidents which are exclusively based on the parent's information and are not objective and accurate, especially during the first year of infant's life, the proposed integrated system offers the medical professionals the opportunity to assess the incidents having recorded bio-signals.
- The use of innovative machine learning algorithms on the multimodal medical signals shall significantly aid the detection with novel quantitative biomarkers of the relevant diseases.
- The medical database implemented shall significantly contribute to research and further mstudies on such early childhood disorders. The system is expected to effectively notify the parents and enable the doctors to identify specific pathologies.

10.5.1 IoT-BASED HEALTH CARE IN INDIA

In the past few years, the health-care landscape in India has seen a scanty growth, as the country is lagging behind in most of the major health indicators. At present, India is responsible for 20% of the burden of diseases across the globe as it has registered alarming newborn mortality rate figured at 40 deaths for a 1000 born alive and a maternal death rate of 100 and 74 for each 100,000 born alive. The adoption of IoT can be conducive in the context of health care in India, only by facilitating with all associated support and factors imperative for implementation of IoT in health care. The need of the day is to produce a body of talented health experts, engineers, technocrats, scientific researchers, and usability experts in design, all uniting to accomplish better health condition of people. Owing to the pervading use of cellular phones, digital communication at present could outreach to even the remote villages, thereby connecting the health-care setups. Existing IoT-enabled devices are being used to capture and monitor health data even in rural areas. To list a few are the fitness bands, smartwatches, heart rhythm detectors, and monitoring patches.

We are in the midst of a global crisis that has come in the form of COVID-19. The ongoing universal challenge due to the COVID-19 outbreak has beaten all boundaries of nationality or race. Health-care services empowered with IoT can come to the rescue of COVID-19 victims, as it administers monitoring of patients over an interconnected network of medical devices and sensors. The technology shall be an aid to boost satisfaction and lower the rate of readmission of patients in the hospital. IoT is a highly promising technology-based support to combat with COVID-19 and can attain remarkable challenges during the lockdown period. It is very helpful to acquire real-time information about the health parameters of the infected person. With a mission to defeat the disease by creating more and more awareness of the civilians about the COVID-19 pandemic, the Government of India has instituted a digital mobile application, namely, Aarogya Setu. The app is aimed at establishing a connection between health-care services and the people of India. The app built by the National Informatics Centre, Govt. of India with its initial release on 2nd April 2020, began to supplement the efforts to track the infection and limit the spread of COVID-19. It was instructed by enabling Bluetooth-based contact tracing, mapping of probable COVID-19 hotspots and publicizing appropriate information. The app could enter 100 million users club as of May 2020 and surpassed any other *Contact Tracing App* in the world. Being available in 12 languages and on Android, iOS, and KaiOS platforms, the app has been used by people countrywide to safeguard themselves and others as well. The app is now regarded by many as their bodyguard. The crucial features of the Aarogya Setu app include transparency, privacy, and security. The product is in compliance with the Open Source Software policy of India. A similar mobile application called as *Close Contact* has been instituted for use of the people in China. This application notifies the user regarding his proximity to a COVID-19 positive person in the neighborhood so that added care can be undertaken soon while roaming outside the residence.

10.6   CONCLUSION

IoT-enabled connectivity in health care is potentially robust to project a health-care setup into a combined, effective, patient-oriented as well as well-organized one. The technology shall assist in cutting down the burden of expensive health-care services through holistic measures and hence, change the prevailing attention of curative care to welfare and belongingness. The case studies reported demonstrate possible smart living solutions based on the integration of IoT and smart biosensors to record, analyze bio-signals non-invasively on the top of an ecosystem. This shall aid in early detection and prevention of potentially hazardous pathological conditions. Health-care services in India have certainly been elevated to the next level when Aarogya Setu app experienced massive use by people to know the proximity to a COVID-19 positive person in the neighborhood, as it administers monitoring of patients over an interconnected network of medical devices and sensors.

IoNT is taking large strides and has multiple potential applications in health care, together with the evolution of novel online applications for diagnosis and medical imaging, stronger pharmaceuticals, equipment for delivery of prescribed drugs, and medical implants. A domain of medicine that utilizes the knowledge of nanotechnology is nanomedicine, which is the outcome of the process comprising (i) diagnosing, (ii) treating, and (iii) preventing diseases as well as traumatic injury by utilizing nanostructured materials and simple nanodevices. The communication of these materials with living systems forms the basis for nanoscale-structured materials and nanodevices as well. Eventually, maybe in the next 10–20 years, nanomedicine is anticipated to offer its exploiting potential in terms of the molecular device and nanorobot an emerging field of technology that creates nanomachine. This apparent technology of nanorobots in medicine marks an emerging era in Nanomedicine. The nanorobots may join the medical workforce through the IoT, benefitting the physicians with the most potent tool to treat severe human diseases. The IoT coupled with Nano Science and Technology promises an unimaginably different world. Any new positive aspects of life would also be associated with a baggage of problems; but, humanity would evolve, alongside, to face the unknown challenges.

REFERENCES

Akyildiz, I. F., Brunetti, F. & Blazquez, C. (2008). Nanonetworks: A new communication paradigm. *Computer Networks*, *52* (12), 2260–2279.

Akyildiz, I. F., & Jornet, J.M. (2010a). The internet of nano-things. *IEEE Wireless Communications*, *17*(6), 58–63. doi: 10.1109/MWC.2010.5675779.

Akyildiz, I. F. & Jornet, J. M. (2010b). Electromagnetic wireless nanosensor networks. *Nano Communication Networks*, *1*(1), 3–19.

Balasubramaniam, S., & Kangasharju, J. (2013). Realizing the internet of nano things: Challenges, solutions, and applications. *Computer: A Publication of the IEEE Computer Society*, *46*(2), 62–68.

Chen, M., et al. (2017). Wearable 2.0: enabling human-cloud integration in next generation healthcare systems. *IEEE Communications Magazine*, *55* (1), 54–61.

Freitas, R.A. (2005). Current status of nanomedicine and medical nanorobotics, *Journal of Computational and Theoretical Nanosciences*, *2*, 1–25.

García-Magariño, I., Lacuesta, R., & Lloret, J. (2018). Agent-based simulation of smart beds with internet-of-things for exploring big data analytics. *IEEE Access*, *6*, 366–379.

Giges, N. S. (2013). Top 5 Trends in Nanotechnology, Retrieved from: https://www.asme.org/engineering-topics/articles/nanotechnology/top-5-trends-in-nanotechnology.

Giovangrandi, L., Inan, O. T., Wiard, R. M., Etemadi, M., & Kovacs, G. T. A. (2011), Ballistocardiography-a method worth revisiting. In *IEEE Engineering in Medicine and Biology Society*, 4279–4282.

Integration of the sustAGE System, Retrieved from: https://www.sustage.eu/?p=1570.

Kailas, A., & Ingram, M. A. (2010). Opportunistic large array-based cooperative transmission strategies. *IEEE Transactions on Wireless Communications*, *9*(8), 2415–2419.

Liu, P., Tao, Z., Lin, Z., Erkip, E., & Panwa, S. (2006). Cooperative wireless communications: a cross-layer approach. *IEEE Wireless Communications*, *13* (4), 84–92.

Lu, D., & Liu, T. (2011). The application of IoT in medical system. *In IEEE International Symposium on IT in Medicine and Education*, 1, 272–275.

Nosratinia, A., & Hunter, T. E. (2004). Cooperative communication in wireless networks. *In IEEE Communications Magazine 2004, 42* (10), 74–80.

Omanović-Mikličanin, E., Maksimović. & Vujović. (2015). The Future of Healthcare: Nanomedicine and Internet of Nano Things. *Folia Med. Fac. Med. Saraeviensis, 50* (1), 23–28.

Pateraki, M., et al. (2020). Biosensors and internet of things in smart healthcare applications: Challenges and opportunities. In *Wearable and Implantable Medical Devices*, ISBN: 978-0-12-815369-7.

Rohokale, V. M., Prasad, N. R., & Prasad, R. (2011). A cooperative internet of things (IoT) for rural healthcare monitoring and control. In *Proceedings of Wireless Vitae 2011, 2nd International Conference on Wireless Communications, Vehicular Technology, Information Theory and Aerospace & Electronic Systems Technology* (p. 49). Chennai, India: River Publishers.

Sadek, A. K., Su, W., & Liu, K. J. R. (2007). Multinode cooperative communications in wireless networks. *IEEE Transactions on Signal Processing, 55*(1), 341–355. doi:10.1109/TSP.2006.885773.

Swan, M. (2014). Nanomedical Cognitive Enhancement, Retrieved from: http://ieet.org/index.php/IEET/more/swan20140711.

The Medical Futurist (2019). A Digitally Caring Environment: The Internet of Things in Hospitals, Retrieved from: https://medicalfuturist.com/iot-in-hospitals.

What's So Special about the Nanoscale, Retrieved from: https://www.nano.gov/nanotech-101/special.

Yang, G., et al. (2014). A health-IoT platform based on the integration of intelligent packaging, unobtrusive biosensor, and intelligent medicine box. *IEEE Transactions on Industrial Informatics, 10* (4), 2180–2191.

11 Machine Learning-Based Smart Health-care Systems

Anil Kumar Swain and Bunil Kumar Balabantaray
NIT Meghalaya
Shillong, India

Jitendra Kumar Rout
KIIT Deemed to be University
Bhubaneswar, India

CONTENTS

11.1 Introduction..195
    11.1.1 Historical Perspective and Trends 196
    11.1.2 Traditional Health Care Vs. Modern Health Care............................196
    11.1.3 Classification of Smart Health Care 196
    11.1.4 Requirements in Smart Health Care..196
        11.1.4.1 Functional Requirements 196
        11.1.4.2 Non-functional Requirements..198
    11.1.5 Smart Health Care Terminologies 198
11.2 Technological Stack in Smart Health Care..199
    11.2.1 Applications of Smart Health Care 201
        11.2.1.1 Medical Imaging 201
        11.2.1.2 Bioinformatics..202
        11.2.1.3 Predictive Analytics 202
11.3 Smart Health Care in COVID-19: A Case Study...204
11.4 Issues & Challenges in Smart Health Care..206
11.5 Conclusion 207
References..207

11.1 INTRODUCTION

It has become very necessary to work on improving our health-care sector. We do now require smart health care as traditional methods are becoming more expensive and time-taking. Nowadays, lots of data are available from health-care sectors and with proper use of those data with appropriate technology, our health-care sector can be improved. Smart health care is primarily concerned with mechanization, technology, analysis square measures that are advancing each day, and historic medication. Smart health care is no longer just a simple technological advancement, rather it's a multilayered associated all-rounder modification. These modifications include changes in medical models (earlier models were disease-centered and currently patient-centered), changes in the construction of knowledge (earlier we had clinical informatization and currently pharmaceutical informatization), substitute in health care management (earlier general management to currently personalized management), and changes in the concept of prevention and treatment (from focusing on disease treatment to focusing on preventive health care). These changes focus on meeting the individual needs of people while improving the efficiency of medical care, which greatly enhances the medical and health service experience, and represent the future development direction of modern medicine (Tian et al., 2019).

11.1.1   HISTORICAL PERSPECTIVE AND TRENDS

There were various developments that took place in the past such as vaccination, antibiotics, and sterilization. Health-care works on three pillars: diagnostic, remedial, and therapeutic (Bhardwaj et al., 2017). Unlike in the past, when doctors did not believe in creating a bond with patients, now doctors try to understand the actual problem with all sorts of diagnosis, before directly going on the point and then follow up with patients to understand if the medication they are given is suitable for them. However with increasing number of patients, doctors are now forced to restrict the time for a patient, and hence modern tools are required to meet the goals of patient-centered care.

11.1.2   TRADITIONAL HEALTH CARE VS. MODERN HEALTH CARE

See Table 11.1.

11.1.3   CLASSIFICATION OF SMART HEALTH CARE

See Figure 11.1.

11.1.4   REQUIREMENTS IN SMART HEALTH CARE

Smart health care is a solution to the traditional health-care system, which is automated and offers solutions to those users who can get comments from various clinics. Smart health care could be divided into various categories such as system management, users, services, applications, connectivity technologies, and medical devices (Figure 11.2).

11.1.4.1   Functional Requirements

Functional requirements are those that are limited to all of the components individually. Consider a smartwatch, which is capable of doing various tasks such as checking

Table 11.1

Difference Between Traditional and Modern Techniques of Health Care

Sl. No.	Traditional Method	Modern Method
1.	Disease-centered treatment model.	Patient-centered treatment model.
2.	Check up as and when required.	On a continuous basis.
3.	Data is disintegrated and occupancy.	Data is integrated, shared, distributed, and is updated on a regular basis.
4.	Health-care professionals were required to take care of patients.	Various IoT and ML-based devices are used to take care of patients.
5.	The appointment is taken by the patient.	The appointment is given to patients by ML-based devices according to the need of disease.
6.	Medical records are maintained on paper.	Medical records are maintained electronically.
7.	It is based on experiences.	It is based on piece of evidence.
8.	It consumes the knowledge available in books.	It learns new actions on the basis of training.

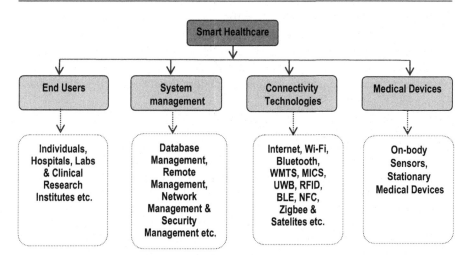

Figure 11.1 Classification of smart health care.

heartbeat, sleep duration, and so on. Here working of each sensor can be a functional requirement. Some functional requirements are given below.

a. **Sensors** such as temperature-measuring sensors, ECG sensors, blood pressure, sensor, SPO2 sensors, motion sensors, EMG sensors, and Gyroscope sensors.

b. **Computing devices** such as smartphones, tablets, PDA's, super computers, and servers.

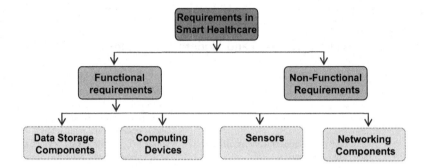

Figure 11.2 Various types of requirements in smart health care.

c. **Data storage components** such as embedded memory of servers, servers that can handle big data, and so on.

d. **Networking components such as** Wi-Fi, Bluetooth, LoWPAN, RFID, and so on.

11.1.4.2 Non-functional Requirements

Non-functional requirements are those basic traits that determine the nature of the health-care system. Some of the non-functional requirements are quality of service, memory, high speed, small form factor, low power, ambient intelligence, connectivity, higher efficiency, reliability, and interoperability in between various platforms.

11.1.5 SMART HEALTH CARE TERMINOLOGIES

a. **Clinical Imaging**

   Clinical imaging is an emerging field in the health-care industry. It helps in image processing and processes a large amount of data more efficiently. It is one of the fastest ways of analyzing and extracting the results from larger data with more accuracy (Miotto et al., 2018). It has helped in the detection and cure of many diseases such as early diagnosis of Alzheimer disease from brain MRIs, automatic segmentation of knee cartilage MRIs to predict the risk of osteoarthritis, manifold of brain MRIs to detect modes of variations in Alzheimer disease. Radiomics is another emerging field that helps in the detection of cancer from medical images (Obinikpo and Kantarci, 2017).

b. **Computer Vision**

   This is one of the greatest successes in deep learning (DL). It focuses on object classification, detection, segmentation by focusing on image and video understanding (Esteva et al., 2019).

c. **Electronic Health Record**

   Electronic health records (EHRs) have provided a larger platform for health care. It is the systematic collection of patient data in digital format. These records are shareable across health-care sectors. It stores all the detailed information of a patient over time. It helps in automatically assigning diagnoses to patients from their clinical reports (Miotto et al., 2018). It also helps in the prediction of suicide risk for the mental health patients using their past medical history as stored in EHRs.

d. **Genomics**

Using genomics in the neural network, conventional machine learning (ML) was replaced with deep architectures, without changing the input features. For example, to predict the splicing activity of individual DNA, fully connected neural networks were used. The model was trained, and the accuracy of the splicing activity was more efficient as compared to the simpler approaches. It could detect the rare mutations in the misregulation of splicing (Miotto et al., 2018).

e. **Health Management**

Health-care management is responsible for organizing and managing the facilities. They also provide administrative knowledge along with business knowledge in terms of health care (Desai & Jhaveri, 2018).

f. **Personalization**

There are various ways to personalize the health-care experience like concierge service, doctor on demand, personal communication with new moms (Orlando Health), Bot to identify the disease with the help of symptoms, and online therapy (Yin et al., 2018).

g. **Smart Equipment**

Smart equipment includes smart asthma monitoring, smart insulin pump, smart drill which drills on the basis of bone density, smart watch, and many more.

h. **Surgery**

Various surgeries can be performed by trained robots, which can eliminate the emergency need of doctors and large number of operation theater.

11.2 TECHNOLOGICAL STACK IN SMART HEALTH CARE

There are many technologies involved in IoT-based health-care system that makes it smart. IoT helps professionals to continuously see patients safely. It has sensors that are used to track the location of various equipment or get the reading of various monitoring equipment. Most applications in the health-care system include the following technologies (Figure 11.3).

i. ML algorithms
ii. DL algorithms
iii. Artificial intelligence (AI)-based smart health care

i. **ML algorithms**

ML is a technology that is used to gain an understanding of the current data and then predict the outcome of future data. There are various models (linear regression, logistic regression, K-NN, decision tree, ensemble

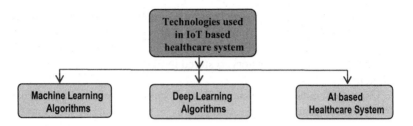

Figure 11.3 Various technologies used in smart health care.

model, and clusters) of ML that can generate the outcome, but the major task is to find the best possible model. There are three types of learning models in ML, namely, supervised, unsupervised, and reinforcement model. In supervised learning models, we know the target variable and perform the analysis on the training data set. Once we have made the model, we can apply the model to a test data set to check the accuracy. For example, let us suppose we have a data set of admission and demographics of the patient; we can easily predict the outcome of 30-day readmission using this technology (Dua et al., 2014). Overfitting is a condition when the model works perfectly on trained data but not on test data, whereas underfitting is a condition when the model works poorly on both trained and test data.

ML uses supervised or unsupervised algorithms to discover patterns in the data to generate the operations needed (Desai & Jhaveri, 2018). The use of supervised or unsupervised ML algorithms depends on factors related to the structure, volume, and operation needed on the data as well as the use case. Besides selecting the most effective algorithmic rule at the start of the latest development, the R&D skilled have to opt for the proper quantity of knowledge to verify and validate the model. Illegal information, surplus information, deficient information, and an excessive amount of information introduce risk into the machine. The risks are supported by the sort of algorithmic rule employed in the model.

It is terribly necessary for developing employees to know the need for privacy in conjunction with security. In ML, a brand new security risk involves the harmful and mischievous introduction of unwanted information into the machine (Al-Azzam & Alazzam, 2019), which might cause invalid outputs and also put the patient's life in danger. The use of moral hackers, however, will facilitate mitigate the chance of unhealthy information in supervised learning that ultimately protects against unhealthy information also because of the patient's aid records.

ii. **DL algorithms**

DL is a subdivision of ML which uses a huge amount of data for solving complex problems. DL provides a better model as compared to ML. It uses various kinds of data such as structured data, unstructured data, and interconnected data. Artificial neural network (ANN) works as a backbone for DL as it also uses big data in its process. Using DL, larger data is analyzed efficiently and rapidly without compromising accuracy (Al-Garadi et al., 2020).

iii. **AI-based smart health care**

In health care, the key challenges are quantity, choice, and speed. There are many large applications and services that need the storage of patient data such as how many times the patient visits hospital and takes services, all this information need to be updated. Currently, with the increase usage of smart devices, social networks, and web services facilitate generation of huge amount of data (Solanas et al., 2017) and to process such data conventional databases and information storage mechanisms are inadequate. To get rid of this, a combination of non-relational and relational databases can be used to store clinical data in digital form. All these data have to be consistent, and databases capable of processing semi-structured are highly essential. Cloud computing's scalability makes on-demand services accessible for a large number of users. Cloud-assistive treatment help medicos offer services to patients irrespective of geographical locations. The combination of Big data with cloud computing and AI techniques proved to even provide better analysis and results.

11.2.1 APPLICATIONS OF SMART HEALTH CARE

The major benefit of smart health care is the analysis and management of the huge generated data. Modification in an already existing application will be helpful. It is easy to access, fast, reliable, and secure. The quality of the health-care system has improved by adding the algorithms of ML and AI. Sensors equipped with smartphones have emerged to be one of the most useful things for the patient to keep track of health monitoring handy. Wireless body area network is the basic component of health-care applications. They are used for patient monitoring tasks. ECG and heart monitoring have become cost-effective in smart health care. Diagnosis is medical imaging, which is one of the important applications of smart health care. Smart health care has also provided features like localization and tracking (Joshi et al., 2020). It helps in monitoring all the risk factors and potentially gives patients direct access to their health, preventive measures as well as to manage the ongoing illness.

11.2.1.1 Medical Imaging

ML has helped within the improvement of care by providing correct unwellness detection and recognition from an outsized heterogeneous set of knowledge. Associate in nursing example is that the detection of skin cancer (Esteva et al., 2019). To do this, ML models learn the necessary options associated with skin cancer from an outsized set of medical pictures and use their learning algorithms to observe the presence of the unwellness.

11.2.1.2 Bioinformatics

The applications of cubic centimeters in bioinformatics have seen a recovery at intervals in the diagnosing and treatment of most diseases. Associate in nursing example is cancer identification where deep auto-encoders are used (Miotto et al., 2018), gene expression is taken as input data; factor selection/classification and gene variants mistreatment microarray knowledge sequencing with the help of deep belief networks.

11.2.1.3 Predictive Analytics

ML is a technology that is used to refer to historical data in order to predict future data, hence we can easily predict some of the data.

a.   **Predicting risk of nosocomial Clostridium difficile infection:** ML models can take EHR as data which includes data regarding medicines, procedures, history of patients, staff record, results of the lab work, and admission details of hospital. This data is used to find patients' probability of CDI. This can be used to generate estimates of risk on a daily basis.

b.   **Predicting reservoirs of zoonotic diseases:** ML models take a data set containing information of rodents carrying zoonotic pathogens as input. Using this data set, researchers have come to know the status of reservoirs with good accuracy; moreover, models predicted those species which were harboring more than two pathogens. This has helped a better search, control of vectors, and research of medicines.

c.   **Predicting clinical outcomes in Ebola virus disease (EVD):** Even though ML techniques are applied on a large data set, here it was applied on a small data set from the cohort. It was easier to predict the outcome of Ebola with just a few symptoms and laboratory results (Qayyum et al., 2020).

d.   **Predicting patients at greater risk of developing septic shock:** In sepsis, it is better to identify at the starting point as later it gives the septic shock. Researchers used the MIMIC - II data set and predicted the patients who will likely receive septic shock with high accuracy.

e.   **Emergency medicine:** There are some tools of CAD that are used to follow laboratory work. These practices have helped to understand the capability to boost health-care quality. Current research studies in this field are mainly concerned to use a wider variety of diseases with more information. The application of ML in most of the cardiovascular-based CAD tools suffers from increased false-positive rates, which helps to detect disease much earlier. The summary of various applications of smart health-care is given in Table 11.2.

Table 11.2
Various Applications of Smart Health Care

Sl. No	Application	Problem	Machine Learning/Deep Learning Technique	Accuracy on Test Data
1.	Medical Imaging	Neural cells Classification	CNN	91.3%
		3D brain reconstruction	Deep CNN	96.49%
		Brain tissue classification	DBN	91.6%
		Tumor detection	CNN	99.12%
		Alzheimer's diagnosis	DNN	84.6%
2.	Bioinformatics	Breast cancer diagnosis	Deep Autoencoder	98.59%
		Gene classification	DBN	Between 65% and 94%. Highest : 94% where feature size = 14
		Protein splicing	DBN	Pan's human PPI data set: 98.78%, Matthew's correlation coefficient (MCC): 97.57%, *Escherichia coli* data set: 95.949%, Drosophila data set: 98.349%, and Caenorhabditis elegans data set: 98.669%. Mus Musculus:92.43%
3.	Predictive Analytics	Disease prediction and analysis	Autoencoder and multilayer perceptron	SVM: 63. 3237, KNN : 66.8479, Random Forest : 55.3202, Proposed method 96.475
			CNN	More than 65%

11.3   SMART HEALTH CARE IN COVID-19: A CASE STUDY

The CIoMT technology which is a subset of IoT, the modern method of treatment helps us to control COVID-19. In this technology the analysis, controlling, and capturing is carried out on a daily basis. Some of the uses are listed below (Figure 11.4).

a.   **Tracking in real time**
     The worldwide daily update in COVID-19 cases along with the amount of cured patients, toll of deaths, and the number of active cases in varied locations typically tracked. As a result, the severity of the illness area unit typically sculptured and conjointly the illness activity is commonly expected mistreatment AI for higher deciding and readiness for management by the health authorities and policymakers (Dhillon & Singh, 2019). The govt. initiatives, health care preventative measures, and treatment procedure updates area unit typically accessible to everybody connected to the CIoMT network.

b.   **Remote patient monitoring**
     As the COVID-19 is extraordinarily contagious, the doctors and health workers are in danger of this infection. The CIoMT helps the doctors to look at the patients' health condition remotely with tip medical data such

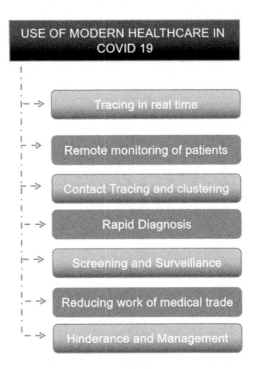

Figure 11.4   Use of smart health care in COVID-19.

as sign level, hexose level, heart bit rate, ECG (EEG), pulse rate, temperature, rate, and so on. Some clinical data acquisition is also possible through wearable IoT devices (Esteva et al., 2019). Since all medical units of the COVID-19 hospitals are extensively connected through the Internet, communication of medical data between different units is easily and quickly accessible. CIoMT is also useful to keep track of old as well having repeat symptoms (Sundaravadivel et al., 2017).

c. **Contact tracing and clustering**

To control the unfolding of the pandemic, the communication tracing of the reported cases is very important, and this repetitive work is commonly minimized if the history of the COVID positive patient is easily accessible within the data that can be accessed by the care authorities. Area-wise cluster and categorization of regions as containment zones, buffer zone, red zone, orange zone, green zone, and so on. Area units are typically rapidly updated depending on the number of confirmed cases through CIoMT. A wise type of positive case area unit is usually collected in periods once the area unit of the medical and health care units is interconnected via IoT. This information and alert for health checkups for the affected house can be accessed by the government, and this could be done chop-chop through the AI framework (Qayyum et al., 2020). The zone cluster put together allows the general public authorities to implement various laws and orders on imprisonment and social distancing.

d. **Rapid diagnosis**

The suspected patients including migrants are isolated although they do not show any symptoms, therefore the speedy identification of those cases is essential. CIoMT allows such persons with travel history to attach themselves to medical services for speedy identification with minimal error. The laboratory technicians will remotely take the X-ray or CT-scan from the management room through the live video streaming which might be processed by AI-enabled visual sensors, therefore taking less time to diagnose and ensure the case. This additionally allows contactless and early detection of the virus (Sundaravadivel et al., 2017).

e. **Screening and surveillance**

The thermal imaging-based biometric authentication info at varied entry points of airports, bus stops, railway stations, malls, hostels, and so on is accessed through CIoMT by the overall public and spotlight authorities for screening and investigating functions (Desai & Jhaveri, 2018). This automatic investigating of the suspected and positive cases can facilitate the unfold of the infection (Sundaravadivel et al., 2017).

f. **Reducing work of the medical trade**

Starting from identifying, monitoring, and medication, the CIoMT helps the restricted attention professionals to touch upon an oversized mass of the population. The CIoMT permits remote viewing of illness that to boot reduces the work. The laboratory technicians needn't visit doorsteps for identification. The AI-infused with IoT sensory data to boot helps in modeling and foretelling of the infection (Chen et al., 2017). Also, the hospitals can guarantee timely doorstep delivery of consultation through telemedicine and delivery of medication by collaborating with block chain corporations.

g. **Hindrances and management**
The unfold of the virus is controlled by the timely intervention of the public and health-care authorities to boost alertness at individual level (Chen et al., 2019). The CIoMT permits one to grasp the positive case inside the neighborhood and to be alert by exploitation sure apps (e.g. Arogya Setu used in India).

11.4   ISSUES & CHALLENGES IN SMART HEALTH CARE

Despite the use of different ML & DL architecture, several challenges are being faced in the health-care sector (Newaz et al., 2019, Chang et al., 2019, Devendran et al., 2018, Mahanty et al., 2018). The challenges in health-care sector are broadly classified as follows (Figure 11.5).

- **Data volume:** DL uses a variety and a wide range of data. Due to the wide range of data, it becomes difficult for the designed model to understand the disease and its variability (Miotto et al., 2018).
- **Data quality:** Health care information contains mixed varieties of information like reports, a laboratory takes a look at samples, patients visit list and details might can (Miotto et al., 2018) contain numerous errors or may even be incomplete. Training to an honest DL model with such an enormous and type of information set is difficult and desires to contemplate many problems such as information meagerness, redundancy, and missing values.
- **Temporality:** The diseases forever increase and alter their nature with time planning a model on a short-lived basis is straightforward however not economical and the planning model mistreatment of the heterogeneous, complicated data set in line with dynamic things may be a troublesome task.

Figure 11.5   Various challenges involved in smart health care.

- **Domain complexity:** The problems in biomedicine related to health care are quite complicated. They need proper research and planning. The diseases are quite heterogeneous, and we don't have complete knowledge about them, their progress, and their cure.

There are also many other features that are promising for the future in DL such as feature enrichment, model privacy, temporal modeling, incorporating expert knowledge, and so on.

11.5   CONCLUSION

The health-care sector is on the verge of a major shift. We are now increasing electronic data of health care, on which when ML is applied we can visualize risks better. ML does not identify the relationships which are not present in the data and hence we do require traditional ways also. However, in the future, interdisciplinary teams may work together and eliminate this traditional way also. DL can operate or dogma to arrange every hypothesis-driven analysis and investigation within the clinical domain that supports all together completely different sources of information. It will open the simplest way toward next generation health care system with a lot of improved options. Now, health care is one in all the growing sectors, and therefore, the quantity of cash spent on this sector is additionally immense, and so we have a tendency to currently need trendy ways in which health care with every kind of automation with the help of advanced technologies.

REFERENCES

Adi, E., Anwar, A., Baig, Z., & Zeadally, S. (2020). Machine learning and data analytics for the IoT. *Neural Computing and Applications*, 32, 1–29.

Al-Azzam, M., & Alazzam, M. B. (2019). Smart city and smart-health framework, challenges and opportunities. *International Journal of Advanced Computer Science and Applications*, 10(2), 171–176.

Al-Garadi, M. A., Mohamed, A., Al-Ali, A., Du, X., Ali, I., & Guizani, M. (2020). A survey of machine and deep learning methods for internet of things (IoT) security. *IEEE Communications Surveys & Tutorials*, 22, 1646–1685.

Bhardwaj, R., Nambiar, A. R., & Dutta, D. (2017, July). A study of machine learning in healthcare. In *2017 IEEE 41st Annual Computer Software and Applications Conference (COMPSAC)* (Vol. 2, pp. 236–241). IEEE.

Chang, V., Cao, Y., Li, T., Shi, Y., & Baudier, P. (2019, May). Smart healthcare and ethical issues. In *1st International Conference on Finance, Economics, Management and IT Business* (pp. 53–59). SciTePress.

Chen, M., Hao, Y., Hwang, K., Wang, L., & Wang, L. (2017). Disease prediction by machine learning over big data from healthcare communities. *IEEE Access*, 5, 8869–8879.

Chen, P. H. C., Liu, Y., & Peng, L. (2019). How to develop machine learning models for healthcare. *Nature Materials*, 18(5), 410.

Desai, A. M., & Jhaveri, R. H. (2018). The role of machine learning in Internet-of-Things (IoT) research: A review. *International Journal of Computer Applications*, 179(27), 36–44.

Devendran, T., Agnes Archana, D. A., & Suseela, S. (2018). Challenges and issues of healthcare in internet of things (IoT). *International Journal of Latest Trends in Engineering and Technology*, Special Issue, April-2018, pp. 86–91. e-ISSN: 2278-621X.

Dhillon, A., & Singh, A. (2019). Machine learning in healthcare data analysis: A survey. *Journal of Biology and Today's World*, 8(6), 1–10.

Dua, S., Acharya, U. R., & Dua, P. (Eds.). (2014). *Machine Learning in Healthcare Informatics* (Vol. 56). Berlin: Springer.

Esteva, A., Robicquet, A., Ramsundar, B., Kuleshov, V., DePristo, M., Chou, K., … & Dean, J. (2019). A guide to deep learning in healthcare. *Nature Medicine*, 25(1), 24–29.

Joshi, A. M., Shukla, U. P., & Mohanty, S. P. (2020). Smart healthcare for diabetes during COVID-19. *IEEE Consumer Electronics Magazine*, 10(1), 66–71.

Mahanty, A., Singh, G., Som, S., & Khatri, S. K. (2018, August). Security issues and challenges in perception layer of smart healthcare. In *2018 7th International Conference on Reliability, Infocom Technologies and Optimization (Trends and Future Directions) (ICRITO)* (pp. 639–644). IEEE.

Miotto, R., Wang, F., Wang, S., Jiang, X., & Dudley, J. T. (2018). Deep learning for healthcare: Review, opportunities and challenges. *Briefings in Bioinformatics*, 19(6), 1236–1246.

Newaz, A. I., Sikder, A. K., Rahman, M. A., & Uluagac, A. S. (2019, October). Health-guard: A machine learning-based security framework for smart healthcare systems. In *2019 Sixth International Conference on Social Networks Analysis, Management and Security (SNAMS)* (pp. 389–396). IEEE.

Obinikpo, A. A., & Kantarci, B. (2017). Big sensed data meets deep learning for smarter health care in smart cities. *Journal of Sensor and Actuator Networks*, 6(4), 26.

Qayyum, A., Qadir, J., Bilal, M., & Al-Fuqaha, A. (2020). Secure and robust machine learning for healthcare: A survey. arXiv preprint arXiv:2001.08103.

Solanas, A., Casino, F., Batista, E., & Rallo, R. (2017, September). Trends and challenges in smart healthcare research: A journey from data to wisdom. In *2017 IEEE 3rd International Forum on Research and Technologies for Society and Industry (RTSI)* (pp. 1–6). IEEE.

Sundaravadivel, P., Kougianos, E., Mohanty, S. P., & Ganapathiraju, M. K. (2017). Everything you wanted to know about smart health care: Evaluating the different technologies and components of the Internet of Things for better health. *IEEE Consumer Electronics Magazine*, 7(1), 18–28.

Tian, S., Yang, W., Le Grange, J. M., Wang, P., Huang, W., & Ye, Z. (2019). Smart healthcare: Making medical care more intelligent. *Global Health Journal*, 3(3), 62–65.

Yin, H., Akmandor, A. O., Mosenia, A., & Jha, N. K. (2018). Smart healthcare. *Foundations and Trends R in Electronic Design Automation*, 12, 1–67.

12 An Exhaustive Survey of Privacy and Security Based on IoT Networks

Santosh Kumar Sahu and Durga Prasad Mohapatra
National Institute of Technology
Rourkela, India

Dibya Ranjan Barik
Oil and Natural Gas Corporation Limited
Dehradun, India

CONTENTS

12.1 Introduction...209
    12.1.1 IoT-based Networks and Related Security Issues 210
    12.1.2 Objective of Threat Detection 211
12.2 Literature Survey 212
12.3 Architecture of IoT 213
12.4 Threat Modelling 216
12.5 Design and Security Challenges in IoT 219
    12.5.1 Design Framework 220
    12.5.2 Countermeasures 222
12.6 Summary..224
References...224

12.1 INTRODUCTION

The invention of Internet of Things (IoT) has taken the human race altogether into a new parlance. It is successfully making our surrounding world smarter and more responsive. This has happened due to cheap computer chips, ubiquity of wireless networks, and the adoption of IPv6. Although IoT was primarily intended for businesses and manufacturing, but now it spreads across sectors such as smart home & office devices, health care, transport, education, government applications, smart cities & planning, water management, smart farming, traffic management, and environment management, to name a few. IoT devices communicate with each other in (near) real time, sharing critical data, thus helping in taking informed decisions. This machine-to-machine (M2M) communication has provided unprecedented opportunities for humankind. The advantages of the M2M communication are given below:

- Accessing faraway information is easy in real time
- Easy communication between connected devices
- Increasing automation, efficiency of services

Like two sides of a coin, IoTs also come with their inherent problems such as little/no security, privacy breach, and so on. Almost every physical object can be converted into a smart object by injecting a chip and connected into the Internet. With this view, many companies are racing to push intelligent devices into market without proper security testing or putting little/no effort on how to secure personal data in these smart devices, just to make them commercially competitive. The disadvantages of M2M communications are discussed below:

- Security compromise
- Privacy Breach
- Over–reliance on technology lead to catastrophe

Due to restriction with low computation power, memory, bandwidth and battery resource, it's not feasible for the devices to execute computationally intensive and latency-sensitive security measures. Hence, complex and robust security mechanisms cannot be accompanied with these devices. In addition, these devices operate in heterogeneous networks, so a common anomaly detection method is difficult to develop to cover all range of IoT devices, making them vulnerable to various types of attacks such as spoofing, disruption of services (DoS/DdoS), privacy breach, device hijacking, and so on. Inside the Smart Home: IoT Device Threats and Attack Scenarios.. For example, there could be so many smart IoT devices present in a smart home such as smart lock, smart refrigerator, smart toy, smart bulb, smart speakers, smart coffee maker, and so on. If a hacker has gained access to any of the devices, he or she may look for vulnerabilities to exploit maliciously IoT security: How these unusual attacks could undermine industrial systems.. Even in industrial application, vulnerabilities can be exploited in smart industrial devices to cause potential sabotage, robot hijacking, and limit the legitimate users by launching DDoS attacks.

12.1.1 IoT-BASED NETWORKS AND RELATED SECURITY ISSUES

As per the forecast made by Gartner Says 5.8 Billion Enterprise and Automotive IoT Endpoints Will Be in Use in 2020, the growth of enterprise and automotive IoT market increase to 5.8 billion endpoints, with a rate of 21% from previous year as given in Table 12.1.

This emphasises the crucial role of IoT and related technologies, but like a double-edged sword, it suffers tremendous privacy and security-related problems. The researcher faces various security challenges such as device authentication, encryption in data communication, privacy, DDoS, probe, U2R and R2L attacks. As we

Table 12.1

Segment Wise IoT Endpoints (Unit: Billion)

Segment	Year (2020)
Utilities	1.37
Government	0.7
Building automation	0.44
Physical security	1.09
Manufacturing & Natural resources	0.49
Automotive	0.47
Health-care providers	0.36
Retail & Wholesale trade	0.44
Information	0.37
Transportation	0.08
Total	5.81

know, the IoT devices have limited CPU and memory capacity executed on battery-operated power supply. To secure the communication, we can't use a strong encryption/decryption approach. The degree of vulnerabilities is less; as a result, the hackers may exploit these devices for malicious intentions. Therefore, as per the limited resources, the alternative approaches such as device registration, MAC filtration, role-based authentication, OTP-based multifactor authentication techniques are used to safeguard the IoT devices.

12.1.2 OBJECTIVE OF THREAT DETECTION

Due to the technological advances, the intruders take its advantages to find new vulnerabilities and get opportunities to exploit the security for their own gain. To counter the newly discovered vulnerabilities, tactics, techniques, and procedures are detailed analysed by the security personnel and develop the solutions or patches that avoid the security breaches. This is a continuous approach like two sides of a coin. To safeguard the digital resources, continuous threat monitoring by intrusion detection system/intrusion prevention systems should be used and whenever a threat discovered, the administrator should take proactive approach to patch to fix the bug or vulnerability of the system. The objective of this chapter is to discuss various threats, their designing approaches, modelling, and countermeasure the intrusive efforts.

The rest of the chapter is organised as follows. Section 12.2 reviews the existing works based on security and privacy issues of IoT networks. Section 12.3 presents different architectures of IoT networks. The details of threat modelling to detect and countermeasures different attacks on IoT networks are discussed in Section 12.4. The design and security challenges in IoT are discussed in Section 12.5. The future scope and direction on security and privacy issues are discussed in Section 12.5.2. Conclusion of the study is given in Section 12.6.

12.2 LITERATURE SURVEY

We have extensively reviewed some existing research works based on security and privacy issues associated with IoT networks as discussed below.
Stephen and Arockiam (2017) suggested an IDS to detect sinkhole attack particularly in the networks, which employs RPL routing algorithm. IDS Agent calculates a
ratio between the number of packets received and the number of packets transmitted.
This ratio is called intrusion ratio. Based upon this IR value, IDS identifies malicious node and alerts other leaf node to isolate it. The purpose of their approach is to
minimise the IR value. Malicious node either drops packets or selectively forwards
packets. Once IR is skewed, malicious node is identified. Raza et al. (2013) implemented a real-time IDS called SVELTE primarily detecting routing attacks such as
sinkhole, altered information, spoofing and selective forwarding. The system integrates a 6LoWPAN mapper, intrusion detection module, and a firewall. Mapped data
is thoroughly analysed to detect intrusion in the network. However SVELTE's true
positive rate was not 100%.

Shreenivas et al. (2017) have worked on SVELTE technique and added another
module called expected transmission metric (ETX) to identify malicious activities
in the network. ETX is a reliable metric; nevertheless, attacks can be done in ETX-
based networks. Hence, they employed geographical hint technique to detect malicious node conducting attacks on ETX networks. The performance of SVELTE
combining rank and ETX module was far better than only SVELTE as true-positive
rate was increasing, despite the higher number of nodes in the network. Pongle
and Chavan (2015) proposed a distributed and centralised approach to detect wormhole attack and attacker. Their proposed system takes into account location information about node and its neighbour information to identify the Wormhole attack
and uses received signal strength to identify the attacker node. It is made especially
for resource-constraint environment, as it is energy-efficient and requires few UD
packets to track the attack. Jun and Chi (2014) proposed an event processing IDS,
where complex event processing (CEP) technology has been incorporated to detect
real-time threats from the events data collected from IoT devices. Because of CEP
large volume of events, data could be processed with low latency.

Summerville et al. (2015) proposed an approach for ultra-lightweight deep packet
anomaly detection. Although it runs on resource-constraint IoT devices, it provides
good differentiation between normal and abnormal payloads. Flowing data bytes in
sequences are called bit-pattern. Overlapping tuples of bytes are called n-grams. Feature extraction happens on these n-grams. Processing network bytes as n-grams, instead of bytes, making it cumbersome for the attacker to evade. Midi et al. (2017) proposed a knowledge driver, self-adaptable expert real-time IDS. It is protocol-agnostic
and can run in any IoT without much performance impact. It first gathers knowledge
about the monitored network and entities and then dynamically configures its real-
time detection strategy. It has a knowledge-sharing mechanism, which facilitates for
collaborative incident detection.

Thanigaivelan et al. (2016) proposed distributed anomaly detection system. The
system is based on the idea that each node monitors packets coming from neighbours.

Upon detection abnormal behaviour, node blocks the packets coming from a rogue node and reports to its parent through a control message called Distress Propagation Object. This message goes on from child to parent until it reaches the edge router node. This message is integrated into the routing protocol in such a manner that it can be used in lossy and low-power networks effectively. Oh et al. (2014) proposed a lightweight IDS. It is an improvised version of the conventional pattern detection security algorithm. They have used a technique called auxiliary shift value to skip large amount of data, which are not required to be matched. When patterns have identical prefix values, upon character matching, matching operation is early terminated. It reduces the memory usage, making this technique suitable for resource-constraint IoT devices. Ioulianou et al. (2018) proposed a lightweight hybrid signature-based IDS. Signature-based IDS are more useful in detecting known attacks and less computational too. Detection module runs on the centralised router and other lightweight modules are deployed in the distributed network. This proposed model is tested against DoS attack like Hello flooding and version number modification.

Prabavathy et al. (2018) proposed a technique based on Fog Computing using Extreme Learning Machine. Distributed fog nodes detect attack at a faster rate in IoT devices. This distribution mechanism is flexible, scalable, and interoperable. Fog computing operates in reduced latency and lower bandwidth arena. While ELM is fast learning technique for single-layer feedforward neural network. It facilitates attack detection in the distributed cloud nodes and final summarisation happens in centralised cloud server. It has been tested against attacks such as DoS, R2L, U2R, and the result shows that it is a highly accurate system with reduced false alarm rate. Some of the related popular IDS approaches for IoT networks are given in Table 12.2. Similarly, the broad attack categories against IoT networks are tabulated in Table 12.3.

12.3 ARCHITECTURE OF IoT

Due to the popularity and outstanding opportunities of IoT devices, more organisations are interested in developing IoT-enabled products in their business processes. The organisations establish their IoT architecture for implementing the products as per their convenience. In this chapter, there are five IoT architectures, as shown in Figure 12.1.

Broadly, architecture of IoT consists of three layers based upon their functions, namely, sensor and actuator layer (information acquisition), network layer (information transmission), and information data-processing and application layer In-Depth view of 4 IoT Architecture Layers.; iot.

Sensor and actuator layer consists of various types of sensors, actuators, and gateways, called edge devices. They acquire various physical parameters, sensory information, or other intended information, that is, gas sensor, proximity sensor, IR sensor, smoke detector, GPS, camera, and so on. Due to commercial viability, these devices needs to be cost-effective, low-power consuming, and smaller in size. Consequently, these devices have low processing power and memory. As a result, they

Table 12.2
Summary of Existing Works on IDS for IoT Networks (Anthi et al. 2019)

Paper	Year	Threat	Methodology	Platform
Stephen and Arockiam (2017)	2017	Sinkhole attack	Validating intrusion ratio (i.e. number of packets received/number of packets transmitted)	-
Raza et al. (2013)	2013	Routing attacks on RPL Protocol	SVELTE IDS using ETX metric and geographical hints	Simulation
Shreenivas et al. (2017)	2017			Simulation
Pongle and Chavan (2015)	2015	Wormhole attack	Location information of node and neighbour information to detect threat and received signal strength to detect attacker	Simulation
Jun and Chi (2014)	2014	Intrusions	Complex event processing technology for complex pattern identification and real time data processing	-
Summerville et al. (2015)	2015	Worm propagation, code injection, Superfluous, Directory traversal	Ultra lightweight deep packet anomaly detection	Empirical (sensor and actuator)
Midi et al. (2017)	2017	ICMP Flood attack, Smurf attack	Self-adapting knowledge driven. It can be network based, hybrid signature/anomaly based, hybrid centralised/distributed	Empirical
Thanigaivelan et al. (2016)	2016	Packet flooding, selective forwarding, clone attack	Hybrid internal anomaly detection	Simulated
Oh et al. (2014)	2014	Routing attacks	Signature based pattern matching engine with auxiliary shifting and early decision	Empirical
Ioulianou et al. (2018)	2018	DoS attack variants viz. Hello flooding, version number modification	Hybrid(centralised/distributed) signature based IDS	Simulated
Prabavathy et al. (2018)	2018	Probe attack, DoS, R2L, U2R	Distributed Fog computing using Online Sequential Extreme Learning Machine	Empirical

Table 12.3
Description of Different Attacks in IoT Networks (Chatterjee et al. 2020)

Name of the Attack	Description
Man in middle attack	Anonymous entity intrudes in the communication between sender and receiver and access confidential information
Sniffer attack	Capturing data during transmission stage in network. If not properly encrypted, hacker gets the sensitive information
IP spoofing	Spoofed source IP address is input to data packet headers to hijack user's browser connected over network
DoS attack	Illegitimate users blocking legitimate users from accessing any device or services by flooding false incoming requests to that particular device or services
Malware attack	Piece of software performing malicious activities on users machines without their knowledge. E.g. Spyware, Ransomware
Sinkhole attack	Rogue node attacks neighbouring node during data transmission to reduce their efficiency
Wormhole attack	Rogue nodes in IoT network creates a tunnel tricking other nodes believe that these two rogue nodes are near to each other and hence attracting data packets. This creates problem in routing and rogue nodes can temper with the data packets.
Black hole attack	Rogue node received data packets from neighbouring nodes and drops them

lack a strong encryption mechanism or complex data securing algorithm. Further, since these devices operate in various environments, providing physical security is cumbersome. Attacks such as tampering, eavesdropping, Radio Frequency jamming, cloning, spoofing, DDoS, and so on always loom over these devices.

Network layer facilitates transmission of collected data, by the sensors and actuators, to cloud, through the Internet or various wireless networks. Attacks like Man in the Middle attack, DoS, DDoS, session hijacking, Sybil, selective forwarding, Hello attack could happen in this layer. Finally, the data processing and application layer consist of cloud technologies to store vast amount of data collected and process them in real-time/near real-time fashion by applying a plethora of analytical algorithms. Delivery of application specific services to users happens through Human Application Interface, for example, smart fire fighting system, smart toy, smart house equipment, smart manufacturing, and so on. This layer is also susceptible to attacks like malicious code injection, cross site scripting, SQL injection, phishing, social engineering, DDoS, and so on.

Although three-layer approach shows the backbone of IoT architecture, a few other models show delved down aspects. One of them is five layer approach shown in Figure 12.1 (Sethi and Sarangi 2017). The objects layer represent in perception layer in a three-layered approach. Application layer remains the same. The remaining layers are as given below:

- Object abstraction: The main objective is to gather data from perception layer and forwards to service management layer/ processing layer and vice

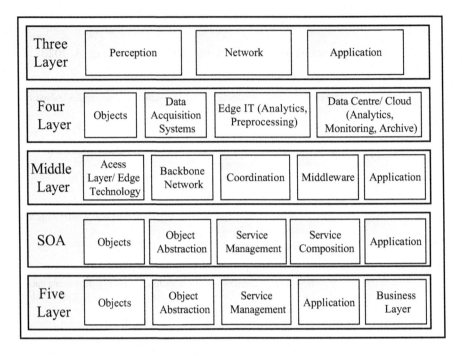

Figure 12.1   A typical IoT architecture.

versa. It employs various communication networks such as NFC, 3G, LTE, Bluetooth, RFID, and so on.

- Service management/processing layer: The objective is to store humongous amount of data, process them and analyse them. This layer has rich set of services for its lower layer. It employs technologies such as database, Big data, cloud technology, and so on.
- Business layer: This layer has offerings to end users. It maintains policies and data privacy.

12.4 THREAT MODELLING

A brief overview of threat of each layer is shown in Figure 12.2. The details of threat, their countermeasures and protocol used in that layer are discussed in Table 12.4.

Cyber Analysts use many techniques/ models to assess attacks. Few of them are discussed below.

- Attack graph
- Kill Chain
- Diamond Model
- Attack surface

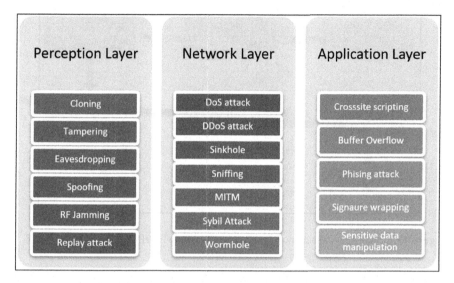

Figure 12.2 Different types of attacks related to IoT architecture/design. (Chatterjee et al. 2020.)

Table 12.4

Threat and Their Countermeasures in Each Layer of IoT Networks

Layer	Threats	Countermeasures	Protocols Used
Perception	Spoofing, Jamming, DDoS, Tempering, Eavesdropping, etc.	Lightweight encryption, Intrusion detection, Physical safety, Password based access, Hash algorithm to conform data integrity, etc.	GPS, CDMA, 3G, 4G, LTE, NFC,etc.
Network	Sinkhole, Wormhole, Cybil, MITM, Session Hijacking, DoS, DDoS, etc.	Lightweight encryption, Intrusion detection, Private & Public key to maintain data integrity, Routing control mechanism, etc.	TCP, UDP, IPv6, RPL, 6LoWPAN
Application	Phishing, Social engineering, SQL injection, Cross site insertion, etc.	Firewall, intrusion detection, encryption, Identity Management, etc.	HTTP(S), XMPP, DSS, MQTT, etc.

We will have a discourse on Kill Chain Modelling to understand possible threat patterns and hence could formulate mitigating strategies as elaborated in Figure 12.3. Delving down the steps of Kill Chain model would illustrate defensive, investigative, and offensive tactics to handle the advanced persistent threats (APTs) and other cyberattacks. For example, understanding the working of a Trojan would prepare the security team to deter/minimise likelihood of attack/loss.

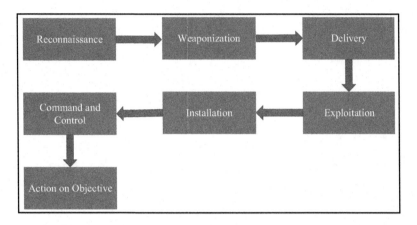

Figure 12.3 Kill chain attack modelling steps. (Kiwia et al. 2018.)

Step 1: Reconnaissance: - Perpetrator tries to gather as much possible data as he/she can from various sources about his/her targets. Some of the sources are free public internet, social profiles of the targets, from other cyber-attacks etc.

Step 2: Weaponization: - Perpetrator readies the malicious payload and embeds within document like PDF or MS Office docs. This payload now has to be sent to target machine or network. Due cognizance has been given in this step to reduce possibilities of attack detection or lessening possibilities of forensic investigations. Weaponization is achieved through

- Host-based evasion
- Network- based evasion
- Anti-Forensic techniques

Step 3: Delivery: Perpetrator finds a way to deliver the payload to target machine/network. Possible measures are

- Email Attachments
- Drive by Download
- Social Engineering

Step 4: Exploitation: This is the phase when actually misuse or exploitation happens. The malicious payload exploits the system vulnerabilities and establishes foothold in the target. Successful exploitation could be

1. Steal credentials
2. Inject malicious code into web applications
3. Steam Cookies
4. Ex-filtration of classified information
5. Downloading other malicious modules
6. Logging Keystrokes

Step 5: Installation: Installing the payload in target machine automatically or by the victim. Perpetrator controls this application and tries to expand reach to other nodes in the target network without making a fuss.

Step 6: Command and Control: Now perpetrator has got strong access to victim machine's resources and exploits them through various control ways such as ICMP, DNS, website, and so on. Through data gathering tools perpetrator gathers classified information, credentials from the target machine.

Step 7: Action on Objectives: Once attack has been done, next step is to ex-filtrate sensitive information about the victim's machine, creating backdoors, encrypts victim's machine data, system disruption, etc.

12.5   DESIGN AND SECURITY CHALLENGES IN IoT

IoTs are developed by many industrial players, leads to diverse structure of their operations and communications. It is really cumbersome to provide end-to-end security in IoT network. Few majors design challenges and their underlying security threats are shown in Figure 12.4. The details of each security threat as discussed in Basu et al. (2015), Fink et al. (2015), Kraijak and Tuwanut (2015), Vasilomanolakis et al. (2015), Suo et al. (2012), Mahmoud et al. (2015) are given below:

- **Heterogeneity and interoperability:** Solution-oriented design of IoT, multiple versions and working with many other vendors' product has led

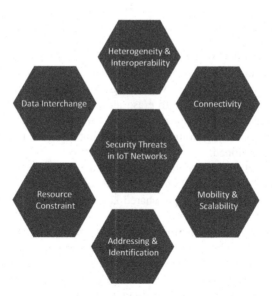

Figure 12.4   Security threats on IoT based networks.

to exposure of its internal working mechanism, APIs, and other resources. These can be misused in a field environment and give rise to spoofing, DOS attacks, MITM attacks, impersonation, and so on.

- **Connectivity:** Peripheral devices in IoT network communicate with each other over low power medium such as NFC, ZIgBee, Bluetooth, and so on. Bridging devices are there to convert these data to suitable for IP network and thus could be possibly act as source of MITM attacks.
- **Mobility and scalability:** Although IoT devices are mobile, yet they come with their inherent disadvantages of frequent disruption in communication or rogue devices latching in place of the actual one.
- **Addressing and identification:** For last mile connectivity, the coordinator node distributes local addresses to the connected local nodes. In general, address allocation does not follow a common standard and hence difficulties in isolating a rogue node from outside network. Moreover, rogue node spoof within the network to impersonate as a router or other node.
- **Resource constraint:** IoT devices lack computation power, memory capacities, network bandwidth, energy, and so on. Devices that are in field can be cloned and tampered unhindered. This leads to data privacy issue.
- **Data interchange:** IoT devices communicate with each other of different origin and components. Hence encrypted data one device, gets decrypted and suitably encrypted as per other devices at the junctions. This process requires sharing of keys and somehow opens up the probability of resource exhaustion and information leakage.

12.5.1 DESIGN FRAMEWORK

We discuss a few design frameworks for overcoming the above security threats.

- **End point physical security:** Devices should not be easily tampered or cloned. Tamper resistance hardware should be used to make IoT. The network should have the resiliency to isolate the tampered node to protect others. Facility should be there to erase all critical information hold by the tempered node remotely to avoid information misuse. Device memory and other critical operating information should not be accessible from outside to avoid being cloned.
- **Bootstrapping and setup security:** During the booting phase, when a node tries to latch up with the IoT network, all of its critical information like cryptographic information, pre-shared keys, private keys, and so on must not be compromised or stolen.
- **Authentication, authorisation and accounting:** Proper authentication mechanism should be in place for any node before communicating with the server. Authorisation mechanism should be there for the legitimate use of resources and privileges. Every activity must be logged for non-repudiation.
- **Data transmission and storage security:** Lightweight cryptic algorithms should be applied before data transmission over network. Node gathered

data should be properly signed and encrypted and sent to huge storage capacities.

- **Network and routing security:** Once data is out of the IoT network and in transmission stage in actual Internet, it also requires proper encryption and integrity checks to avoid eavesdropping and distortion attacks. In addition to that, intrusion prevention and rollback mechanism should be in place to reduce further damage.
- **Multilayer Security:** Security should be there in protocol stack , which operates in lossy network as well as which operates in Ethernet network.

These are a few frameworks for providing security to IoT devices and the data, although there are many other solutions cropping up each day. Security has become a great concern for IoT, as the focus is on making these devices marketable rather than strengthening internal security. The growth of IoT is always creating a safe haven for cyber criminals. Vast protocols created for IoT are being misused by rogue users to obfuscate their malicious usage trails. Many of them are making huge fortune by selling malwares, botnets, hacking kits, providing hacking as a service in cyber world. Poor monitoring of these devices, obsolescence of embedded OS within devices, lack of accountability are making these devices vulnerable to cyberattacks. SCADA systems, ICS which are connected to IoT also come under threa. For example, 2008 Explosion in highly secure Turkish pipeline. In cyber world, security stands on the triad CIA, that is, Confidentiality, Integrity and Availability. So is IoT. Let us discuss first the general security challenges and then challenges pertaining to each layer of IoT. General security challenges could technological or design principle-based. Technological challenges are like underlying wireless technology, energy requirement and dissipation by IoT, scalability, and so on. While design principle challenges are like ensuring confidentiality, accountability, integrity, end-to-end security, and so on.

Few points worth considering during the security design of IoT framework:

- IoT operating software should be authorised
- While booting up and connecting to network, devices should authenticate themselves
- Since devices have limited resources, firewall or IDS to filter incoming/outgoing packets to/from these devices
- Regular updates/patches should be installed

Data should be confidential throughout its life cycle and must be available to its intended users, processes, or machines. Proper data management mechanisms ought to be applied to protect data from internal or external malignant players. Whole IoT network Is based on the principle of data exchanges. Therefore, while data is sent/received/during transmission, end-to-end security must be provided to ensure its integrity. IoT devices or claimed services or the data should be accessible to intended users in timely fashion, it is availability principle for IoT. Mutual authentication is required between IoT devices at the starting of each interaction so that each device must identify and authenticate every other interacting device. All protocols or algorithms intended for IoT should be lightweight, keeping in mind resource-constraint nature of the devices. IoT devices are vary in nature, underlying hardware, software, protocols, working mechanism, what not. It is imperative for each device to be compatible (key management, encryption techniques, operating protocols, etc.) with every other interacting device to overcome heterogeneity challenges. Policies defined for current computers and networks may not be enforceable to IoT, due to their typical nature. Therefore proper policies, SLA must be defined for IoT to enforce standardisation and this will help in creating trust among users and growth of IoT usages.

Maintaining privacy of user data is major concern for IoT network starting from data gathered in the device stage, storage stage, communication stage, processing stage. So concealing sensitive personal information is a prime challenge for IoT devices. Few challenges are discussed below:

- **Privacy in device:** IoT devices should be robust and reliable. Else vulnerabilities would be exploited to leak out classified information.
- **Privacy during communication:** To safeguard data during transmission phase proper encryption algorithms to be used.
- **Privacy in storage:** Sensitive information should be stored safely using encryption.
- **Privacy at processing:** While processing classified information should be treated with due diligence. Digital Right Management should be used to control sensitive information processing and protect against illegal usage and re-distribution.

12.5.2   COUNTERMEASURES

Term is organised defence, which is a strong blow to the offensive and malicious intent. Both manufacturers and vendors should be partner in designing security aspects of the devices. Devices should be thoroughly tested particularly all security aspects to maintain CIA of data so the vendors. Data is safe only when CIA are intact throughout its life cycle. Security requirement can be broadly divided into five parts, namely, network security, identity management, privacy, trust, and resilience as shown in Figure 12.5.

- **Network security:** Sub-components of network security to be taken care of for the security of the data. IoT is full of heterogeneous devices. Communication between any two devices requires confidentiality. Integrity should be maintained to confirm no modification or loss of data undetected. Authenticity confirms only authenticated devices take part in communication. Availability confirms devices or services should be available to intended parties. Data confidentiality protects the data from unauthorised access and tampering efforts on the data. To countermeasure such efforts in the dynamic network topology of the IoT network, key exchange methods are used to protect the data. To ensure the integrity during data transmission,

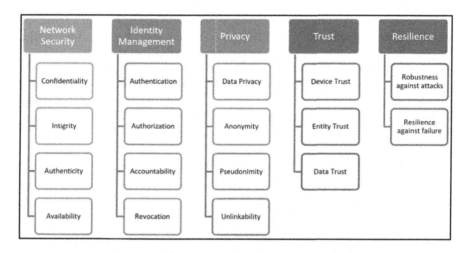

Figure 12.5 Security requirement and their sub-component. (Vasilomanolakis et al. 2015.)

role-based access control and two-phase authentication methods should be adopted. To ensure the availability, it becomes utmost essential to safeguard the network from DDoS attacks.

- **Identity management:** In the humongous IoT network, the relationship between services, services, users, and owners is intricate. Hence, identities should be properly managed through various processes like authentication, authorisation, accountability, and revocation of privileges. It is important to assign proper authorisation to authorised service or equipment; at the same time, it is also imperative to revoke privileges from the same entity once de-authorised or role changed.
- **Privacy:** Privacy remains the most prominent topic in IoT security. Data privacy deals with anonymisation of data in order to prevent a person's identity from being exposed. Personal identifiable information should be stored safely to avoid leakage of data privacy.
- **Trust:** IoT network depends upon trusted devices, entities, and data to maintain communication quality.
- **Resilience:** It is highly improbably to stop all attacks on IoT. Instead, devices should be robust to overcome attacks and design should be fail proof. Recovery mechanism and fail over should be provided to maintain IoT operations under attack situation. Devices should close all unused ports and deactivating unnecessary services to keep themselves away from malicious usage. Incase a device is infected, it must have robust mechanism to isolate itself from the rest of the network or it must intimate the firewall about its probable failing. Upon any offensive event detected, responsive action must be triggered to prevent the malicious use of the device. In this way, a device could be resilient against failure.

These are major security countermeasures, but not limited to. One major push for security enhancement in IoT should come from regulatory authorities. Their guidelines enforce a level of transparency among users. Regulations will make sure manufactures must upgrade and secure these devices with the firmware upgrade, security patches, and so on. Few other improvements are like designing of lightweight cryptographic solutions, applying Machine Learning algorithm to maintain security in IoT network, federated architecture, spreading awareness among IoT users, and so on.

12.6 SUMMARY

In this chapter, we discussed how security plays a vital role in IoT networks. The hackers attempt to compromise the devices and take control of them and create nuisance without knowing in-depth knowledge due to ready-made platforms available for security assessment and penetration testing toolkits. It is of utmost essential to gain prior knowledge on the IoT architecture, the possible attacks on layers, the countermeasure techniques used for threat modelling to safeguard the IoT devices before development. The architectures used for design IoT applications are briefly elaborated. The security issues faced during the deployment and how to avoid the intrusive efforts, the threat modelling concepts are elaborated to study different attacks on various layers. The possible attacks widely used to compromise the security of IoT networks are explained. The design principles and security challenges and countermeasures of the security threat are also briefly explained in this study.

REFERENCES

Anthi, E., Williams, L., Słowińska, M., Theodorakopoulos, G., and Burnap, P. (2019). A supervised intrusion detection system for smart home iot devices. *IEEE Internet of Things Journal*, 6(5):9042–9053.

Basu, S. S., Tripathy, S., and Chowdhury, A. R. (2015). Design challenges and security issues in the internet of things. In *2015 IEEE Region 10 Symposium*, pages 90–93. IEEE.

Chatterjee, J., Das, A., Ghosh, S., Das, M. K., and Bag, R. (2020). 8 a review of cyber attack analysis and security aspect of iot-enabled technologies. *IoT: Security and Privacy Paradigm*, page 159.

Fink, G. A., Zarzhitsky, D. V., Carroll, T. E., and Farquhar, E. D. (2015). Security and privacy grand challenges for the internet of things. In *2015 International Conference on Collaboration Technologies and Systems (CTS)*, pages 27–34. IEEE.

Gartner Says 5.8 Billion Enterprise and Automotive IoT Endpoints Will Be in Use in 2020. Website:https://tinyurl.com/y67zb7da (Accessed on 20/08/2020).

In-Depth view of 4 IoT Architecture Layers. Project Website: https://www.hiotron.com/iot-architecture-layers/ (Accessed on 25/08/2020).

Inside the Smart Home: IoT Device Threats and Attack Scenarios. Website:https://tinyurl.com/y8zjaq5a (Accessed on 20/08/2020).

Internet of things (iot) connected devices installed base worldwide from 2015 to 2025. https://www.statista.com/statistics/471264/iot-number-of-connected-devices-worldwide/. Accessed: 15-07-2019.

IoT security: How these unusual attacks could undermine industrial systems. Website:https://tinyurl.com/y6w2fdn5 (Accessed on 20/08/2020).

Ioulianou, P., Vasilakis, V., Moscholios, I., and Logothetis, M. (2018). A signature-based intrusion detection system for the internet of things. *Information and Communication Technology Form.*

Jun, C. and Chi, C. (2014). Design of complex event-processing ids in internet of things. In *2014 Sixth International Conference on Measuring Technology and Mechatronics Automation*, pages 226–229. IEEE.

Kiwia, D., Dehghantanha, A., Choo, K.-K. R., and Slaughter, J. (2018). A cyber kill chain based taxonomy of banking trojans for evolutionary computational intelligence. *Journal of Computational Science*, 27:394–409.

Kraijak, S. and Tuwanut, P. (2015). A survey on iot architectures, protocols, applications, security, privacy, real-world implementation and future trends.

Mahmoud, R., Yousuf, T., Aloul, F., and Zualkernan, I. (2015). Internet of things (iot) security: Current status, challenges and prospective measures. In *2015 10th International Conference for Internet Technology and Secured Transactions (ICITST)*, pages 336–341. IEEE.

Midi, D., Rullo, A., Mudgerikar, A., and Bertino, E. (2017). Kalis–a system for knowledge-driven adaptable intrusion detection for the internet of things. In *2017 IEEE 37th International Conference on Distributed Computing Systems (ICDCS)*, pages 656–666. IEEE.

Oh, D., Kim, D., and Ro, W. W. (2014). A malicious pattern detection engine for embedded security systems in the internet of things. *Sensors*, 14(12):24188–24211.

Pongle, P. and Chavan, G. (2015). Real time intrusion and wormhole attack detection in internet of things. *International Journal of Computer Applications*, 121(9):1–9.

Prabavathy, S., Sundarakantham, K., and Shalinie, S. M. (2018). Design of cognitive fog computing for intrusion detection in internet of things. *Journal of Communications and Networks*, 20(3):291–298.

Raza, S., Wallgren, L., and Voigt, T. (2013). Svelte: Real-time intrusion detection in the internet of things. *Ad Hoc Networks*, 11(8):2661–2674.

Sethi, P. and Sarangi, S. R. (2017). Internet of things: architectures, protocols, and applications. *Journal of Electrical and Computer Engineering*, 2017:1–25.

Shreenivas, D., Raza, S., and Voigt, T. (2017). Intrusion detection in the rpl-connected 6lowpan networks. In *Proceedings of the 3rd ACM International Workshop on IoT Privacy, Trust, and Security*, pages 31–38.

Stephen, R. and Arockiam, L. (2017). Intrusion detection system to detect sinkhole attack on rpl protocol in internet of things. *International Journal of Electrical Electronics and Computer Science*, 4(4):16–20.

Summerville, D. H., Zach, K. M., and Chen, Y. (2015). Ultra-lightweight deep packet anomaly detection for internet of things devices. In *2015 IEEE 34th International Performance Computing and Communications Conference (IPCCC)*, pages 1–8. IEEE.

Suo, H., Wan, J., Zou, C., and Liu, J. (2012). Security in the internet of things: a review. In *2012 International Conference on Computer Science and Electronics Engineering*, volume 3, pages 648–651. IEEE.

Thanigaivelan, N. K., Nigussie, E., Kanth, R. K., Virtanen, S., and Isoaho, J. (2016). Distributed internal anomaly detection system for internet-of-things. In *2016 13th IEEE Annual Consumer Communications & Networking Conference (CCNC)*, pages 319–320. IEEE.

Vasilomanolakis, E., Daubert, J., Luthra, M., Gazis, V., Wiesmaier, A., and Kikiras, P. (2015). On the security and privacy of internet of things architectures and systems. In *2015 International Workshop on Secure Internet of Things (SIoT)*, pages 49–57. IEEE.

13 Distributed Denial-of-Service Attacks in IoT Using Botnet: Recent Trends and Challenges

Padmalaya Nayak and Surbhi Gupta
GRIET
Hyderabad, India

Pallavi Shree
Amity University
Patna, India

CONTENTS

13.1 Introduction 228
13.2 DDoS Attack: Background Discussions 229
    13.2.1 Types of DDoS Attack 230
    13.2.2 DDoS Attack Employing Botnet 232
    13.2.3 Types of Botnet 232
        13.2.3.1 Botnet C&C Systems 232
        13.2.3.2 Botnet Architectural Model 234
        13.2.3.3 Botnet Life cycle 235
        13.2.3.4 Botnet Communication Mechanism 235
13.3 IoT Botnets 238
    13.3.1 Traditional Botnets vs IoT Botnets 238
    13.3.2 Evolution of IoT Botnets 238
    13.3.3 Example of IoT Botnets 239
13.4 IoT Botnets Detection Approach 241
    13.4.1 IoT Botnet Detection Techniques 242
13.5 Conclusion and Future Work 244
References 244

13.1 INTRODUCTION

In recent years, the Internet of Things (IoT) has witnessed significant growth and billions of devices are connected in 2020 as per the prediction of Gartner about the IoT (Nayak, 2016). Owing to the enormous growth of IoT, hackers are eager to execute different types of attacks by exploiting the computational and communication capabilities of IoT devices. There are many types of cyberattacks such as denial-of-service (DoS), distributed denial-of-service (DDoS), password attacks, birthday attacks, phishing attacks, man in the middle attacks, and so on. (Hoque et al., 2014; Bhattacharyya and Kalita, 2013), which target the information-processing systems, computer networks, infrastructures or personal devices using various methods to steal, modify, and destroy the data. Today, particularly, DDoS attack is the major concern on the Internet, and it occurs in networks with the interconnected systems. These networks consist of several computers and IoT devices that are infected by malware. These devices are remotely controlled by the attacker, and each device is called a bot or zombie. The group of bots or zombies are called a botnet, which is controlled by an attacker or botmaster. The botmaster is used to perform different activities such as DDoS attacks, spamming, click fraud, phishing attacks, and so on. Millions of IoT devices are interconnected through the Internet and causing significant risks to Internet security. The major concern is that IoT devices have several vulnerabilities to be utilized to create IoT botnets. These devices are large in number and create a huge risk to Internet security compared to ordinary Internet-connected devices. In addition, the resource limitations and protocol diversity make IoT devices a valuable target for attackers to create IoT botnet. Security issues of these devices are not emphasized by the manufacturers due to large-scale deployments. Many of the devices are operated with a fixed key permanently. Most of the time, devices are manufactured in bulk, carry default usernames and passwords. Further, all the devices carry the same default username and password. As these devices are not assembled with built-in security features, they are more prone to attack. Therefore, security is the prime concern in IoT devices due to the large growth of Internet-connected networks. Botnet formation attack is one of the most dangerous attacks, which spreads very quickly and impacts prominently. In this context, DDoS attacks stemming from IoT botnets are a key issue in today's Internet that requires urgent attention. The main intention of this attack is to block the network resources and make them unavailable to genuine users. For instance, as public users are permitted to create secret channels publicly or privately in Internet relay chat (IRC) networks, an online chatting system between the attacker and the agent commenced communicating using IRC channels (Lu and Ghorbani, 2008). An IRC-based DDoS attack model and agent-handler DDoS attack model are almost similar except that instead of exploiting a handler program executed on a network server, an IRC server monitors the addresses of the connected agents and facilitates the communication between them (Ma et al., 2010). The DDoS attack is not limited to a specific application, it covers almost all the layers of the transmission control protocol (TCP)/IP protocol stack. Various types of DDoS attacks (Hoque et al., 2015; Nayak et al., 2017) exist on the Internet that the Internet provider and the user must be aware of it. Figure 13.1. Illustrates the example of a DDoS attack.

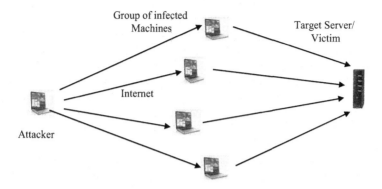

Figure 13.1   Example of DDoS attack.

The rest of the chapter is organized as follows. Section 13.2 gives a brief history of DDoS attacks and the origin of botnets. Section 13.3 explains different types of botnets and their features in detail. Section 13.4 presents various types of botnets and their detection mechanisms in detail. Section 13.5 deals with the future work followed by the concluding remark.

13.2   DDoS ATTACK: BACKGROUND DISCUSSIONS

Denial-of-service (DoS) attack is a tool used by cybercriminals to prevent a net-worked system from providing services or information to legitimate users. This type of attack is mainly carried out to hamper the reputation and business of the vic-tim networks. Originally, these attacks were carried out by a single source called a "Denial-of-Service" or (DoS) attack. DOS attacks are easier to locate and miti-gate. However, nowadays, the attackers coordinate an attack that is carried out from different sources controlled by a central point. These are called "Distributed Denial-of-Service" (DDoS) attacks. DDOS attacks have been the biggest threat to the avail-ability and accessibility of any networked resource. DoS attacks have been prevalent for quite some time. However, it came into the picture in 1974, when a program was written by a 13-year-old girl that brought down a learning management system con-nected to a huge number of terminals at the University of Illinois computer learning lab (Dear b, 2010). In 1999, the first attack came to the front that used 200 zombie hosts to bring down the University of Minnesota's network for 2 days. One of the biggest and first takedowns of large companies occurred in the 2000s when Micheal Calce "MafiaBoy" targeted giant corporates such as eBay, Yahoo, Amazon, and so on. Such attacks came to be termed DDoS attacks. As per Kaspersky Lab, attacks have been doubled in the first quarter of 2020 as compared to 2019. Even though detection mechanisms have been improved a lot, these attacks have become more aggressive and futile. Botnets were once very complicated and advanced distributed systems covering millions of computes, but now they are used in a tiny network of a few dozen hosts with centralized control. A botnet, or zombie network, is a network of infected computers carrying a malicious program through which cybercriminals monitor and control the infected computers remotely without the knowledge of the users. The entire group of cybercriminals uses the zombie networks as a source of their income. These sources of income include various attacks such as DDoS attacks, stealing confidential information, spam emails, phishing, SEO spam, click fraud, and distribution of malicious programs. In 2005, Norman Elton and Matt Keel from the College of William & Mary gave the name to bot networks "the single greatest threat facing humanity" (Elton and Keel, 2005). Although it looks like an overestimation of DDoS attacks, it is one of the biggest concerns being faced by the Internet community. As per the Council of Economic Adviser (CEA) of the Government of the United States, malicious cyber activity cost the U.S. economy between \$57 billion and \$109 billion in 2016 (CEA, Govt of USA, 2018). The same is the case in the Indian scenario; it is discussed in *The Economics Times* article, Indian Computer Emergency Response Team (ICERT) that the botnet-infected systems in the country were 25,915 in 2007, which increased to about 6.5 Million in 2012 (*Economics Times*, 2014).

13.2.1 TYPES OF DDoS ATTACK

The primary goal of the DDoS attack is to dissipate the target resources and to establish a denial of service in the network. It occurs in all the layers of the TCP/IP protocol stack. Although many studies show that DDoS attacks are two types such as at application layer and network Layer (Igure and Williams, 2008), we briefly summarize DDoS attacks are four types based on application, protocols, the volume of traffic, and non-planned. These are categorized into application level, protocol-, volume-, and zero day-based attack. We have summarized various types of DDoS attacks and presented them graphically as shown in Figure 13.2.

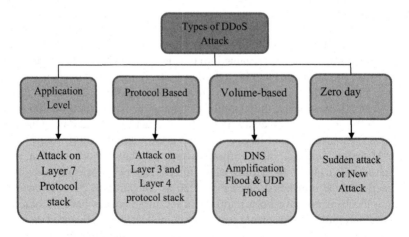

Figure 13.2 Types of DDoS attack.

- **Application-level attack:** The attackers at the application level exploit the weakness of the application layer in the TCP/IP protocol stack. To perform DDoS attacks, the attackers attack the target server by flooding legitimate requests. It establishes a connection to the target machine and exhausts all the server resources by monopolizing the processes and transactions. In network-layer DDoS attacks, the target server or intrusion detection system (IDS) can easily identify genuine packets from DDoS packets, whereas, in the application layer, it is very difficult to distinguish the DDoS traffic and normal traffic. In an application-layer DDoS attack, the attacker employs a flood of legitimate requests to attack the victim server. In this attack model, a TCP connection gets established by any type of zombie machine with the victim server through a genuine IP address. Without a genuine IP address, the TCP connection cannot be established. Examples of this type of attack are HTTP Flood, attack on DNS service, and so on.
- **Protocol-based attack:** In a protocol-based attack, the attacker makes a target by exploiting the weakness on layers 3 and 4 protocols of the ISO/OSI protocol stack. SYN flood attack is an example of a protocol-based attack at the transport layer. Specifically, SYN flood is meant for TCP. In brief, TCP uses three-hand mechanisms to opt for the data transfer from clients to servers. First, an SYN message is sent by the client to the server and gets a reply from the server followed by an acknowledgment from the client. In this attack, the client takes this opportunity and sends a huge number of SYN messages to overwhelm the targeted server's capacity. Ping attack is another example of a protocol attack and targeted for Internet control message protocol (ICMP). It is similar to a SYN flood attack, instead of the SYN messages ICMP echo request messages are flooded in the target network. The UDP flood is another example of a protocol-based attack. An attacker sends a great number of user datagram protocol datagrams to random ports and the port becomes quickly overwhelmed.
- **Volumetric attack:** In a volumetric attack, the invader sends a large amount of data or request packets to the targeted network to exhaust the bandwidth of the network. Typical request sizes are in terms of 100 Gbps. However, recent request sizes have been extended to 1 Tbps. Volumetric attacks are dominant due to other technical barriers for generating high volume traffic, but it is very easy to implement by using a simple amplification technique. Amplification techniques generally originate from many sources like IP addresses or networks. Therefore, it is very difficult to mitigate this type of attack compare to other attacks that originate from a single source.
- **Zero-day attack:** Zero-day DDoS attack is not actually a "zero-day attack" but does exist. This implies that it is not known to the software vendors, but it is known to the attacker. The attackers use different protocols in different ways to launch the attack, which was not done previously. It has happened previously quite a bit with reflection attacks, DNS protocol was used. However, over time, this attack has been leveraged to NTP, SNMP, RIP, and so on. Sometimes, it is quite hazy to think of "zero-day attack" as these protocols exist for a few years.

13.2.2 DDoS ATTACK EMPLOYING BOTNET

A botnet is a method of implementing a DDoS attack. This chapter focuses on the botnet technique, with the major concern being that DDoS attack is more diligent than DoS attack and even more dangerous than DoS attack. It is very important to study botnet technology and analyze its characteristics in detail, predicting the growth of IoT devices in the near future. Botnets are defined as collections of infected computers (bots/zombies) that are controlled by a master remotely called botmaster through a common command and control (C&C) channel. Botmaster controls and communicates with the bots using compromised devices designated as the C&C server. The other name of C&C servers is known as handlers. Botmaster sends the command to all bots through handlers to execute the attack and, in turn, bots perform the attack on target systems. Mostly, the bots are controlled by their architecture, command, and control mechanisms such as IRC, HTTP, P2P-based, or DNS. Most of the time, owners of such zombies are not even aware that their machine has been compromised. These zombies are used to launch an attack. Botnets are involved in various malicious activities such as launching (DDoS) attacks, sending spam and trojan, phishing emails, illegally distributing pirated media and software, stealing information and computing resources, identity theft, and e-business extortion, and so on. It is commonly assumed that approximately 80% of e-mail traffic is spam and usually sent through botnets. In summary, there are four components in a botnet such as (i) attacker, (ii) botnet controller, (iii) infected hosts, and (iv) victim.

13.2.3 TYPES OF BOTNET

Botnets can be broadly classified into three groups, that is, based on C&C mechanism, topological structure, and architectural structure. In the C&C mechanism, the attacker uses the C&C system to play a major role in executing a DDoS attack. There are four types of C&C channels used by the attacker. These are Central, P2P, hybrid, and random. There are also four types of topologies used by hackers, namely, (i) star topology, (ii) ring-star topology, (iii) hierarchical topology, and (iv) topology is not specific. In DDoS attack, mostly three architectural models are used such as (i) agent-handler model, (ii) IRC-based botnet model, and (iii) Web-based model. The graphical presentation of "botnet types" is shown in Figure 13.3.

13.2.3.1 Botnet C&C Systems

Botnet C&C mechanisms are classified into four types as centralized, P2P, random, and hybrid to compute DDoS attacks. Few botnets are static, and few are mobile. Static botnets are several types and have been active for the past few years (Hoque et al., 2015). Agobot, SDbot, RBot, Spybot, Conflicker, Torpig, Strom, Grum, Cutwail, Rustock, and GTbot are examples of Static botnet. Mobile botnets are composed of smartphones, controlled remotely by a botmaster. Every bot has its advantages and limitations. Both the static and mobile botnets operate in four modes as discussed above.

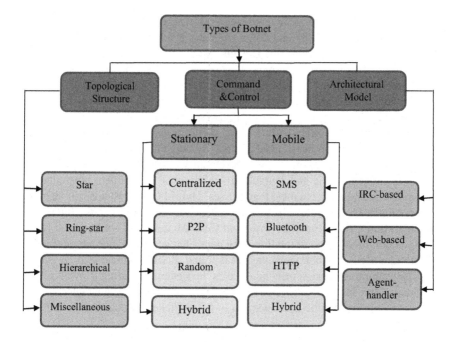

Figure 13.3   Types of botnet.

- **Centralized model:** In centralized model, a single server acts as a botmaster or bot herder. The botmaster controls each bot by using an IRC channel on intermediate devices called C&C server. Communication generally happens in four steps. (i) Once infected, a bot connects to a predefined IRC channel and waits for the instructions. (ii) Botmaster uploads instructions for bot clients on the IRC channel. (iii) All the bots download the commands and execute them. (iv) Bots report the result of the commands that are executed. SDBot, RBot, and Agobot are a few examples of centralized botnets. Botnets based on centralized C&C architecture can be easily mitigated by the attacked host. Even though this type of botnet has several advantages, it has some limitations. If the server fails, then the bots may fail as they will not receive any message.

- **Peer-to-peer model:** Peer-to-peer (P2P) model has gained huge popularity as it has several advantages over centralized model. This model embeds the control structure inside the botnet by eliminating the sole point collapse that appears in centralized botnet architecture. Bots in the P2P model act both as client and command centers. All the bots work together with their neighboring nodes to propagate data. All the list of trusted computers is maintained by all P2P botnets. This list is used to propagate and update malware and other information. There are three types of architectures that are used in P2P. These are unstructured C&C server, structured server C&C server, and

super peer P2P server. P2P has several advantages over the centralized botnet. (i) It is not easy to shut down and there is a high survivability rate. (ii)
It is robust and reliable. (iii) Design complexity is not so high. However,
without a centralized command P2P botnets can be hijacked by someone
else other than the original creator. To prevent such scenarios, these botnets
are encrypted Gameover ZeuS and ZeroAccess botnet, which is an example
of the P2P botnet.

- **Hybrid C&C Model:** Hybrid C&C model is designed by exploiting the
  features of both the centralized and p2p model. It has two types of bots: servant bots and client bots. A servant bot can act as both client and server and
  contains both static and routable IP addresses. But the client bots contain
  dynamic and non-routable IP addresses. Hybrid C&C systems can also be
  configured behind firewalls without the global connectivity to the Internet.
- **Random C&C Model:** In a random C&C Model, each bot knows at least
  one bot. In this type of topology, the controller scans the Internet randomly
  to identify a new bot. Once the bot is found, it sends an encoded message.
  This simplified design principle makes the system appealing. However, the
  major drawbacks of the system are high latency and lack of guaranteed message delivery. Single bot detection is not sufficient to compromise all the
  bots.

13.2.3.2   Botnet Architectural Model

As discussed in Section 13.2.3, there are three basic models used in DDoS attacks
on IoT infrastructure. These botnet models are the agent handler model, IRC-based
botnet model, and Web-based model (Alomari et al., 2012). In this section, we have
demonstrated each model very briefly.

- **Agent handler model:** The agent handler model has four elements; the
  master, the handlers, the agents, and the victim. The master is the main attacker who initiates the game, compromises some systems to control them
  under his administration. The handlers are the second category of players,
  and they are mostly malicious software that resides in the remote machines
  handled by the attacker to launch the attack. The attacker uses the handler
  through a victimized machine and makes the system complicated so that it
  will be difficult to track the attacker. The third category of players, composed of software on a set of infected machines, launch the DDoS attack by
  involving a large no of agents. The attacker exploits the weaknesses of protocols such as TCP, UDP, and ICMP and fulfills his desire. The last player
  is the victim, which may be a single machine or multiple target machines.
- **IRC-based model:** IRC-based DDoS attacks are now most popular as they
  can create huge volumes of attack traffic instantly. In the IRC model, clients
  are connected to the agents through the IRC communication channel. IRC
  is a text-based chat system that allows users to initiate communications
  in channels and causes bots to be produced. The IRC model benefits the

attacker by providing the legitimate IRC ports to send instructions to the agents. The legitimate ports help in tracking the DDoS command packets. Clients in IRC-based DDoS attacks include TCP, UDP, and ICMP protocols.

- **Web-based model:** Despite the huge popularity of the IRC-based method, bots in the Web-based model are used to provide statistics to the website. In a Web-based model, bots simply report the statistics to the website, whereas others are designed and regulated through complex PHP scripts. The encrypted message transfer takes place over 80/443 port and HTTP protocols. The main advantages of a Web-based model are as follows: (i) easy configuration, (ii) consumption of less bandwidth, (iii) improved reporting and command functions, and (iv) resistance to botnet hijacking through chat room hijacking.

13.2.3.3   Botnet Life cycle

Botnet life cycle consists of five phases as initial infection, second-level infection, third-level infection, malicious activity followed by maintenance, and up-gradation (Bailey et al., 2009). A graphical representation is illustrated in Figure 13.4. In the initial phase, the botmaster (mastermind) sends malware to the target machines to infect them. In the second phase, the mastermind tries to log into an IRC server or any other communication medium to set up the botnet through the infected machines. In the third phase, the spammer pays off the owner of the bot for getting access right. In the fourth phase, the spammer gives instructions to send spam or infected codes to many other machines in the victim's network. The fifth phase takes care of maintenance and up-gradation of activities (Figure 13.6).

13.2.3.4   Botnet Communication Mechanism

The communication mechanism is a part of the botnet design lifecycle. The communication mechanisms used between any two bots or between a bot and its C&C server

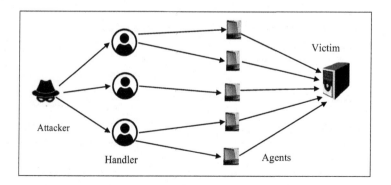

Figure 13.4   Botnet using agent handler model.

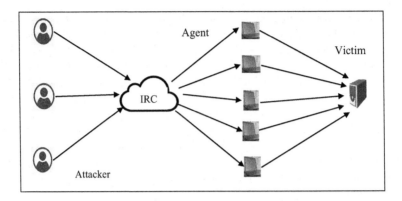

Figure 13.5   Botnet using the IRC-based model.

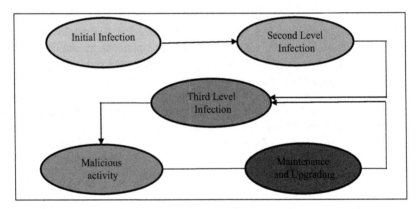

Figure 13.6   Botnet life cycle.

can be categorized into two classes such as push-based mechanism and pull-based mechanism (Wang et al., 2009).

- **Push-based mechanism:** The first-generation botnets use central C&C mechanisms through IRC and other relevant channels. They follow the push-based approach to remain connected with selected channels and others. The other name of push-based communication mechanism is also known as command forwarding. In this type of communication mechanism, initially, the botmaster uses the commands to a few bots only. The bots, who receive the original command from the botmaster, further propagate the commands to other bots. This mechanism considerably eliminates the requirement for the bots to keep checking or requesting new instructions from the botmaster. However, as the traffic is mostly one-way from the server to bots, this can cause an alarm which might result in detection of the botnet. This mechanism is mainly used in IRC botnets. AgoBot, SDBot,

Phatbot, SpyBot, and GTbot are examples of a few botnets, which work on the principle of push-based mechanism. Figure 13.7 shows the example of push-based mechanisms in detail.

- **Pull-based mechanism:** Pull-based communication mechanism is also called Command Publishing/Subscribing used in botnets. Mostly, HTTP botnets are the ones that use this mechanism. In this method, the commands are actively downloaded from a location where the botmaster uploads all commands to be executed as displayed in Figure 13.8. Botnets using pull-based mechanism are BlackEnergy, Festo, Grum, Zeus, SpyEye, Citadel, and TDL-4.

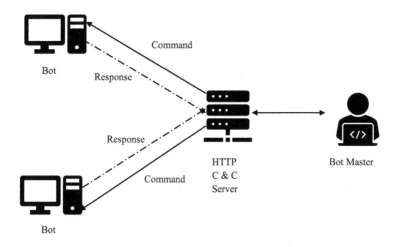

Figure 13.7 Push-based model.

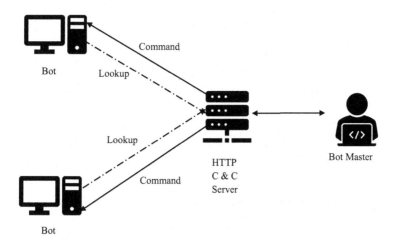

Figure 13.8 Pull based model.

13.3   IoT BOTNETS

Botnets are a collection of devices or computers connected through the internet, infected with malware, and remotely controlled by a malicious actor. These computers are known as zombies. Although botnets are not new in concept, IoT botnets are a collection of smart electronic devices hijacked by cybercriminals with the same objective as a conventional botnet. IoT devices pose similar processing capability to simple computers along with more dedicated functions. These devices are widely available in marketplaces but most of them are not equipped with security features to defend against cyberattacks. Botnets were primarily designed for spam advertized pharmaceuticals, advertisement click fraud, and robbing bank credentials (Khan et al., 2012), while IoT botnets have been discussed to launch DDoS attacks (CYRN Cyber Threat Report, 2017). Based on the CYRN Cyber Threat Report, IoT devices perform DDoS attacks being constantly on-line combining a large number of compromised devices and allow criminals to perform powerful attacks. Mirai is an example of such an IoT botnet. Many variants such as Satori, Okiru, Persirai, Masuta, and Puremasuta are examples of IoT botnets (Kambourakis et al, 2017) as source code was revealed on the Internet. Moreover, botnets behave similarly regardless of their malware family. IoT botnets and regular botnets pose the same botnet life cycles while compromising new devices such as (i) scan the connected devices for open ports, (ii) brute force the exposed ports to gain access to victims, (iii) destroy the similar contenders on the infected machines, (iv) form a C&C channel with the mastermind, (v) sometimes execute or delete malicious script, (vi) search for new instances and spread through the network, (vii) perform various attacks or other malicious actions. In all these attacks, initial compromise remains the same, other actions might differ. For instance, Satori tries to be connected on ports 37 215 and 52 869 (Netlab, 2017) while Persirai exploits directly to access password data (Kolias et al., 2017). However, once the IoT-devices are infected, they can execute DDoS attacks on a large scale due to the in-built vulnerability in IoT devices. For example, Mirai botnet was used to register a 1.1 Tbps botnet attack against the French cloud service provider.

13.3.1   TRADITIONAL BOTNETS VS IoT BOTNETS

There is not much difference between traditional botnet and IoT botnet except the configurations of devices. The difference between traditional botnets and IoT botnets is given in a tabular form as shown in Table 13.1.

13.3.2   EVOLUTION OF IoT BOTNETS

The evolution of botnets that emerged into IoT botnets is discussed in this section referred from (Sha et al., 2018; CYRN Cyber Threat Report, 2017) and illustrated in Table 13.2.

Table 13.1

Comparison of Traditional Botnets and IoT Botnets

Sl. No.	Parameters	Traditional Botnet	IoT Botnet
1.	Architecture	Both centralized as well as P2P	Both centralized and P2P
2.	Life cycle	Infection – Connection with master node – execute the attack – report back results	Same as traditional botnets
3.	Implementation	Difficult – Higher computing power and frequent auto-mated monitoring makes it difficult to implement.	Easy-less processing power of IoT devices makes it easier for IoT botnets to be imple-mented.
4.	Available target	Probable targets are fewer as the number of devices lesser and has greater protection features.	A huge number of probable targets as online IoT devices have grown to a very large scale and limited to no protec-tion features.
5.	Target devices	Computers, Servers	IP Camera, TV, Smart wellness devices, watches and TVs
6.	Detection technique	Easy	Difficult
7.	Detection time	Can be detected quickly by a user as interaction through GUI is more frequent.	Interaction less with IoT devices Web-based GUI. Botnets go unnoticed for longer durations.
8.	Scale of impact	Less	High

13.3.3   EXAMPLE OF IoT BOTNETS

In this section, there are three important botnets such as Kiten, Qbot, and Mirai discussed briefly, and their evolutions, architectural differences are presented in a tabular form as shown in Table 13.3.

- **Kaiten:** Kaiten is the oldest open-source botnet and belongs to the botnet malware family developed in 2001. It is not quite popular among the pub-lic. It communicates through C&C servers and mostly relies on the popu-lar IRC protocol. The infected devices receive the commands from an IRC channel hardcoded within the Kaiten binaries. The other name of Kaiten is Tsunami. Kaiten's script also allows it to work on multiple hardware archi-tectures, making it a relatively versatile tool for cybercriminals. These codes can be compiled over various hardware platforms, namely, SH4, PowerPC, MIPSel, MIPS, and ARM. Besides, recent Kaiten variants are capable of killing competing malware and permitting themselves to fully control a de-vice.
- **Qbot**: Qbot belongs to a very old malware group, but it provides a signif-icant platform for botnet designers. Bashlite, Gafgyt, Lizkebab, and Torlus are well-known names for Qbot. The most important thing about Qbot is

Table 13.2

Development of IoT botnets (Year-wise)

Year	Description
1988	Phone Home is considered the first Internet worm. Robert Morris Jr. released Morris Worm also called "Phone Home".
1999	The trojan is called Sub7 and a worm is known as Pretty Park. Both are used to pay attention to IRC channels.
2004	Phatbot is also called as Agobot which opens a back door on a host and connects it to its own P2P network.
2006	Zeus Trojan acts as a financial service. It was designed to steal banking credentials.
2008	Grum Malware was the third largest botnet in 2012 and was capable to delete billions of messages per day.
2011	Gameover Zeus is a p2p malware, belongs to the Zeus family.
2012	Aidra was described by its author as an "IRC-based mass router scanner exploit".
2013	MiscoSMS was the first android botnet reported by cybersecurity experts.
2014	BASHLITE was launched to compromise IoT devices that operate on Linux.
2016	Remaiten combines the DDoS features of Tsunami and the scanning ability of BASHLITE.
2016	Mirai is the first IoT botnet used in a few devices to infect a large number of devices.
2017	Reaper is one category of IoT botnets that target the victims of known susceptibilities. Particularly, it targets wireless IP-based cameras and other IoT devices using a list of known usernames and passwords.
2018	Double Door is an IoT Botnet uses two backdoor exploits by avoiding firewall and modem security.
2019	Muhstik had exploited CVE-2019-2725, an Oracle WebLogic server.

Table 13.3

Summary of Three Important IoT Bots

Name of the Bots	Year	Features	Type	Hardware Support
Kaiten	2001	Known as Tsunami and recent variants implement a bot-killing feature	C&C Type based on the IRC Protocol	Multiple
Qbot	2008	Known as Bashlite, Gafgyt, Lizkebab, or Torlus and recent variants can uninstall other pieces of botnet malware	Uses C&C servers, but does not rely on IRC. It uses TCP	Multiple
Mirai	2016	Open-source code enables novice attackers to make new, more aggressive variants and source code contains the capability to kill competing processes	Uses C &C Servers and propagates through weakly configured IoT devices	Multiple

that its source code contains only a few files. It is very difficult for new bot-net developers to use it by considering only a few files as discussed in the malware tutorial in cybercriminal forums. The source code of Qbot can also boost several architectures like Kaiten, but the malware depends on TCP instead of IRC to communicate with its C&C servers. Qbot variants are also capable of killing rival malware.

- **Mirai botnets:** Mirai launches the DDoS attack through the target servers by infecting the weakly configured IoT devices. The first Mirai botnet attack was launched on 16th September 2016 by security journalist Brian Krebs through a DDoS attack. Then Mirai botnet source code was leaked after 14 days of attack by "Anna-Senpai" on 30th September 2016. Later, a similar kind of attack known as the largest DDoS attack was targeted at a popular dynamic DNS provider, Dyn. Then, a heavy magnitude attack of 1.2 Tbps was instructed towards Dyn over the DNS port 53, on 21st October 2016. The targeted IoT botnet attack took down several websites. Since then, many variations of the Mirai botnet have appeared to the public. These are listed as Akiru, Katrina_V1, Sora, Saikin, Owari, Josho_V3, and Tokyo, and so on. Almost all the Mirai variants have a similar code structure as they imitate the Mirai source code. Mirai mainly spreads by infecting IoT devices such as routers, Webcams, DVRs that run over some version of BusyBox. Then it takes over the administrative control through brute force attack maintaining a small ID password dictionary. Mirai has become a popular botnet malware unit, producing many variants. The main purpose was to create Mirai to publish a DDoS tool for sale and to make money out of the business. After the public release of the Mirai source code, it became a tricky player for IoT malware. When it first entered the botnet malware domain, it made a name for itself through the attack on Dyn, a Domain Name System (DNS) hosting provider, that disrupted widely used websites and services.

13.4 IoT BOTNETS DETECTION APPROACH

The term "detection" refers to notify the vulnerabilities of connected IoT devices. A detection is a proactive approach without triggering events or user interaction. It targets the first phase of the botnet life cycle, whereas the term "defense" is considered as a reactive approach that is triggered by the output of the detection mechanism. The defense mechanisms are related to configure the firewall rules that isolate IoT devices from the Internet and prevent the devices from becoming part of the botnet. Detection mechanisms can be implemented in mainly two ways. The first method includes techniques that work on a host machine. This type of detection approach mainly focuses on the various processes executed on the host machine and does not perform any analysis on network traffic. On the other hand, another class of detection mechanism focuses on the monitoring of network traffic. However, the approach of analyzing the network traffic can be quite diverse for different techniques. One category of network traffic-based detection mechanism injects their own packets into the network and keeps a record of the behavior of the network towards

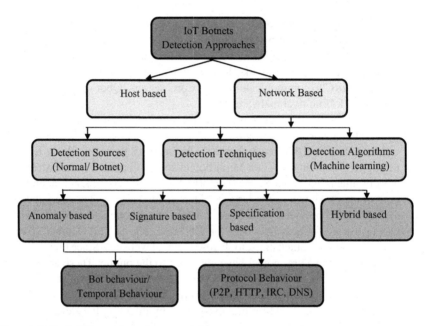

Figure 13.9   IoT botnet detection approach.

those packets. These mechanisms are called active monitoring techniques. There are also a group of techniques where traffic on the network is just supervised for abnormal activities. These are called passive monitoring techniques. We can also classify the network-based detection mechanisms used for detecting malware as illustrated in Figure 13.9. The network-based IoT botnet detection technique is again categorized from three different angles, namely, detection sources, detection techniques, and detection algorithms. Detection sources mean where the packets are captured not that how the packets are captured. For instance, virtual internal networks can verify normal packets from internally controlled packets, whereas real network should capture packets from real traffic. Further, the detection techniques are classified into anomaly-based, signature-based, specification-based, and hybrid-based. It uses the main techniques for detection. The detection algorithms differentiate the types of algorithms used such as supervised, semi-supervised, unsupervised, signal processing, and applying heuristic rules.

13.4.1   IoT BOTNET DETECTION TECHNIQUES

- **Anomaly-based botnet detection technique**
  The main objective of the anomaly-based detection technique is to identify the abnormal behavior in the IoT-based network. To effectively differentiate between an abnormal and normal behavior of the network, this approach first makes a profile of the normal behavior of the IoT-based network. One of the simpler approaches that works on unencrypted traffic is

proposed by Chen and Lin (2015). This method is effective for a botnet of a smaller size. Botnet activities are identified, and abnormal IRC traffic is detected efficiently. Using the network gateways as the collection point for IRC traffic, this method uses a homogenous response and group activity to identify the abnormal activity.

a. Bot Behavior/Temporal Behavior

There are some anomaly-based detection techniques where the interaction or behavior of the bot is analyzed and detected with another bot. It also involves monitoring and analyzing the behaviors of a group of bots acting as a single bot. Temporal behavior means it can detect behavior changes over time. For instance, in the case of a BotMiner, bytes per second and flows per hour are monitored along with an eye over the behavior of the bot getting connected to a large number of SMTP servers and requesting MX records on most of those established connections (Gu et al., 2008a; 2008b).

b. Protocol Behavior

The anomaly-based technique can also focus on the protocols being used or targeted to carry out the activities of the botnet and their corresponding bots. The BotSniffer looks into the traffic for the occurrence of a large number of port scans to detect bots (Gu et al., 2008a). Whereas BotMiner not only checks for port scans but also for undue more frequent connection requests to SMTP servers (Gu et al., 2008b). There are cases like stability (Li et al., 2010) where the focus is on P2P protocols and extremes like N-gram (Lu et al., 2011) where behavior and features are calculated for each protocol separately.

- **Signature-Based Botnet Detection**

These mechanisms require an already existing malware dataset. They use the bot binaries of existing botnets to study the behavior of the bots. It is comparatively easier to implement. However, these tools fail as modern botnets keep their signature updated frequently, and there are several botnets with similar functions but a diverse signature. One of the easiest implementations of the signature-based detection technique (Behal et al., 2010) is to analyze the outbound traffic and implemented it at an educational organization. The traffic was observed, recorded, and then compared with an already stored malware signature. An intrusion detection system is proposed to monitor the networks for already existing malicious activities and policy violations based on matching attack signatures (P. Ioulianou et al., 2018). In their IDS module, they used two main components: one is router IDS and the other one is detector IDS. Router IDS is used to detect the firewall modules. Detector IDS is used to log traffic and report it to router IDS. Thereafter, router IDS is classified based on known attacks.

- **Specification-Based Botnet Detection**

  These mechanisms are similar to the anomaly-based approach with the only difference of considering the specifications of the systems. The authors have proposed an automatic interference technique by taking malware samples and malware binaries for a specification of malware network protocol (Carli et al., 2017). They generated a fingerprint from the combination of malware structure and functions as every malware has its custom binary format and each C&C protocol has its own family. However, most C&C network traffic is encrypted so that they apply dynamic traffic analysis to extract C &C system keys. A detection technique is proposed for IoT botnets during the malware propagation stage where infected devices start to exploit other devices in the network using a brute force attack strategy (Prokofiev et al., 2018).

- **Hybrid-Based Botnet Detection**

  In most cases, the combination of two different approaches fetches the best results because they help in minimizing false positives and false negatives that come up during the process of botnet detection. Synthesis of low energy consumption anomaly and signature-based IoT detection has been proposed in which game theory is used to determine whether or not an IDS agent should initiate anomaly detection (Sedjelmaci et al., 2016). Another implementation fuses anomaly-based and specification-based detection techniques (Bostani and Sheikhan, 2017) where a specification-based detection agent analyzes the host behavior and forwards the results to the root node, which uses an anomaly-based technique to determine the bots.

13.5 CONCLUSION AND FUTURE WORK

IoT networks demand strict security policies to handle communication and connectivity issues effectively since they find difficulties in detecting the attack. Because of this fact, this chapter presents an overview of DDoS attacks on the Internet. Further, the botnet-based DDoS attack for the IoT-based networks are discussed in-depth as botnets are the main facilitators for modern DDoS attack. We have discussed different types of botnets, their communication mechanisms, and different types of detection approaches and methods to protect IoT-based networks. Some of the research issues and challenges are discussed here to push the research further. Even if many detection mechanisms are discussed in the literature, they are restricted to low-rate or high-rate DDoS attacks, not for both. The performance of the network depends on the network conditions and multiple user parameters. Therefore, the development of defense techniques avoiding all these limitations is an important issue for further investigation.

REFERENCES

Alomari, E., Manickam, S., Gupta, B., Karuppayah, S., & Alfaris, R. (2012) "Botnet-based distributed denial of service (DDoS) attacks on web servers: Classification and art," *Int. J. Comput. ppl.*, Vol. 49, no. 7, pp. 24–32.

Bailey, M., Cooke, E., Jahanian, F., Xu, Y., & Karir, M. (2009). "A survey of botnet technology and defenses," In *Proceedingd of the IEEE CATCH*, pp. 299–304.

Behal, S., Brar, A. S., & Kumar, K. (2010). "Signature based botnet detection and prevention," In *Proceedings of International Symposium on Computer Engineering and Technology*, pp. 127–132.

Bhattacharyya, D. K., & Kalita, J. K. (2013). *Network Anomaly Detection: A Machine Learning Perspective, Published by Chapman, and Hall/CRC Press*, Boca Raton, FL.

Bostani, H., & Sheikhan, M. (2017). "Hybrid of anomaly-based and specification-based IDS for Internet of Things using unsupervised OPF based on MapReduce approach." *Comput. Commun.*, vol. 98, pp. 52–71.

Botnet Evolution Infographics (2017) *CYRN Cyber Threat Report*.

Carli, L., Torres, R., Modelo-Howard, G., Tongaonkar, A., & Jha, S. (2017). "Botnet protocol inference in the presence of encrypted traffic." In *IEEE INFOCOM 2017-IEEE Conference on Computer Communications*, pp. 1–9.

Chen, C., & Lin, H. (2015). "Detecting botnet by anomalous traffic." *J. Inf. Secure. Appl.*, vol. 21, pp. 42–51.

Council of Economic Adviser, Govt. of USA. (2018). *The Cost of Malicious Cyber Activity to the U.S. Economy*. https://www.hsdl.org/?abstract&did=808776.

Dear, B. (2010). *PLATO History:* Perhaps the First Denial-of-Service Attack? www.Platohistory. Org. http://www.platohistory.org/blog/2010/02/perhaps-the-first- denial-of-service-attack.html.

Elton, N., & Keel, M. (2005). "Who owns your network? A discussion of bot networks". In *The Security Professionals Conference* (Security 2005).

Kambourakis, G., Kolias, C., and Stavrou, A. (2017). "The mirai botnet and the iot zombie armies," In *Military Communications Conference (MILCOM), MILCOM 2017, IEEE*, pp. 267–272.

Kolias, C., Kambourakis, G., Stavrou, A., & Voas, J. (2017) "DDoS in the IoT: Mirai and other botnets," *Computer*, vol. 50, no. 7, pp. 80–84.

Gu, G., Perdisci, R., Zhang, J., & Lee, W. (2008a). "BotMiner: Clustering analysis of network traffic for protocol- and structure-independent botnet detection", In *SS'08: Proceedings of the 17th Conference on Security Symposium. USENIX Association: Berkeley, CA, USA*, pp.139–154.

Gu, G., Zhang, J., & Lee, W. (2008b). "BotSniffer: Detecting botnet command and control channels in network traffic", In *Proceedings of the 15th Network and Distributed System Security Symposium (NDSS), San Diego, CA*.

Igure, V., & Williams, R. (2008). "Taxonomies of attacks and vulnerabilities in computer systems", *IEEE Commun. Surveys Tuts.*, vol. 10, no. 1, pp. 6–19.

Ioulianou, P.P., Vasilakis, V.G., Moscholios, I., & Logothetis, M. (2018). "A signature-based intrusion detection system for the internet of things", *Paper presented at Information. and Communication Technology Form*, Graz, Austria.

Li, Z., Wang, B., Li, D., Chen, H., Liu, F., & Hu, Z. (2010). "The aggregation and stability analysis of network traffic for structured-P2P-based botnet detection." *J. Networks*, vol. 5, pp. 517–526.

Lu, W., Rammidi, G., & Ghorbani, A. (2011). "Clustering botnet communication traffic-based on n-gram feature selection." *Comput. Commun.*, vol. 34, pp. 502–514.

Lu, W., & Ghorbani, A. A. (2008) "Botnets detection based on IRC community," In *Proceedings of the IEEE GLOBECOM*, pp. 1–5.

Hoque, N., Bhattacharyya, D., & Kalita, J. (2014). "MIFS-ND: A mutual information-based feature selection method," *Expert Syst. Appl.*, vol. 41, no. 14, pp. 6371–6385.

Hoque, N., et al., (2015). "Botnet in DDoS attacks: Trends and challenges" *IEEE Commu. Surv. Tutor.*, vol. 17, no. 4, pp. 2242–2270.

Khan, R., Khan, S.U., Zaheer, R., & Khan, S. (2012). "Future internet: The internet of things architecture, possible applications, and key challenges", In *Proceedings of IEEE 10th International Conference on Frontiers of Information Technology*, pp. 257–260.

Ma, X., et al., (2010), "A novel IRC botnet detection method based on packet size sequence," in *Proceedings of the IEEE ICC*, pp. 1–5.

Netlab, (2017) "Warning: Satori, a Mirai branch is spreading in worm style on port 37215 and 52869," http://blog.netlab.360.com/warning-satoria-new-mirai-variant-is-spreading-in- worm-style-on-port-37215-and-52869-en/.

Nayak, P. (2016). "Internet of things issues, challenges and applications", In *Handbook of Research on Advanced WSN Applications, Protocols and Architecture*, IGI Global ISSN-2327–3305, pp. 353–368.

Nayak, P., Suseela, R.S.U., & Trivedi, V. (2017). "A review on DoS attack on WSNs: Defence and detection mechanisms", In *International Conference on Energy, Communication, Data Analytics and Soft Computing (ICECDS)*, pp. 453–461.

Prokofiev, A.O., Smirnova, Y.S., & Surov, V.A. (2018). "A method to detect Internet of Things botnets", In *2018 IEEE Conference of Russian Young Researchers in Electrical and Electronic Engineering (EIConRus)*, pp. 105–108.

Sha, K., Wei, W., Yang, T.A., Wang, Z., & Shi, W. (2018). "On security challenges and open issues in internet of things", *Futur. Gener. Comput. Syst.*, vol. 83, pp. 1–33.

Sedjelmaci, H., Senouci, S., & Al-Bahri, M. (2016). "A lightweight anomaly detection technique for low-resource IoT devices: A game-theoretic methodology", In *2016 IEEE International Conference on Communications (ICC)*, pp. 1–6.

Wang, P., Wu, L., Aslam, B., & Zou, C. (2009). "A systematic study on peer-to-peer botnets." In *2009 Proceedings of 18th International Conference on Computer Communications and Networks*, pp. 1–8.

14 Detection of Node Cloning Attack in WSN to Secure IoT-based Application

A Systematic Survey

Pinaki Sankar Chatterjee
Kalinga Institute of Industrial Technology
Bhubaneswar, India

CONTENTS

14.1 Introduction...247
    14.1.1 Node Cloning Attack in WSNs 250
14.2 Countermeasures for Node Cloning Attacks 251
    14.2.1 Centralized Techniques 252
    14.2.2 Distributed Techniques..254
14.3 Performance of Node Cloning Attack Detection Schemes.........................256
    14.3.1 Discussion 257
14.4 Conclusion 259
References...259

14.1 INTRODUCTION

There exist a conflict of opinion among different groups of academician, researcher, developer, and corporate people regarding the globally accepted definition of the Internet of Things (IoT). The most suitable definition of IoT could be: "A robust and open network of intelligent objects able to consciously organize, share knowledge, information and data, change the environment and act and react in situations."Over the last decade, the term 'IoT draws the attention of many people by presenting its future goal toward providing a worldwide network of physical objects, which include anything, anytime connectivity for anyone. In short, it could connect things-to-things, human-to-things, and human-to-human in a global network. A unique identity has been assigned to every object in the network. Here, different types of sensors are embedded in the entity called object and are linked with each other, which can generate a huge volume of data for computer analysis.

Traditionally, the radio frequency identification (RFID) and WSN are regarded as the basis of IoT applications. Using RFID, a low-cost identification and tracking can be done. Using WSN, we can achieve more sophisticated sensing in IoT application. The WSN solution has already covered a large range of IoT applications and research. Now the time has arrived where the generic WSN components have to develop in such a way so that it can handle specific IoT application requirements. In WSN, the recent development of wireless communication, analog and digital technology enables low-power and low-cost sensor nodes to communicate themselves and sense an event as a group (Chatterjee and Roy 2015). The set of sensor nodes can perform high-quality sensing through collaborative methods. In the cluster-based networks, these sensor nodes can form different clusters among themselves. There exist one cluster head for each cluster. These nodes can sense some events and can make a decision by themselves or these sensing instances are sent to the central device called base station for a collaborative decision. Each sensor node has a limited processing power. The WSNs communicate with each other when an event occurs through the 2.4 GHz ISM spectrum band (Akan et al. 2009; Chatterjee and Roy 2017). We don't require any license to use ISM band (Chatterjee and Roy 2018a; Chatterjee and Roy 2018b) . Many other technologies such as Bluetooth, Wi-Fi, cognitive wireless sensor networks, and so on are also using this ISM band for data transfer (Haykin 2005; Mitola 2000). WSN has different security threats, which can destroy the effectiveness of its use in IoT applications. It is not possible to fulfil the requirement of IoT applications like low-cost and reliable service unless all the security threats of WSN can be resolved (Chatterjee and Roy 2015).

The various types of security threats of WSN are shown in Figure 14.1.

- The first group is attacks on communication. In this type of attack , the attacker controls data transfers between nodes for a specific purpose. The goal may be to isolate a node in order to attempt to alter the behavior of the entire network.
- The second category of attack is attack on privacy. WSNs make it possible to share resources in order to create contact and to be environmentally conscious. In this type of attack, such access is exploited by the attackers.
- The third category of attack is on targeted node. It is possible to seize a node and use reverse engineering by attackers to become an instrument for mounting such attack.
- The fourth type of attack is the attack on security and privacy policies.
- The fifth kind of attack is an attack on power use. Battery life is a vital consideration for WSN. The attacker can cause sleep torture on an energy-restricted node in this attack by involving in needless communication work to kill its battery life rapidly.
- Finally, in the cryptographic attack, the attacker tries to find the vulnerabilities in the network by analyzing the transmitted information.

In this chapter, we made a comprehensive survey of the node cloning attack, a type of communication attack, and its defensive measures available for WSN.

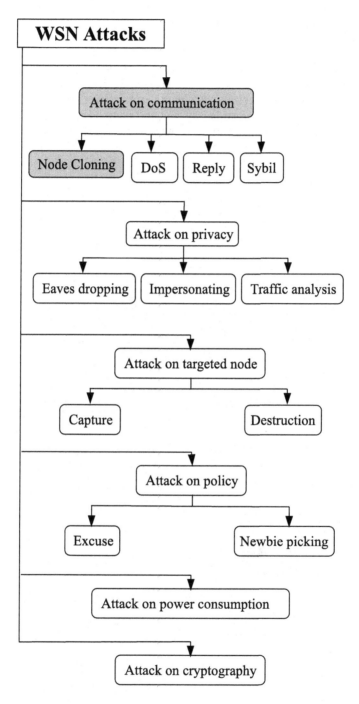

Figure 14.1   Attacks on WSN.

14.1.1   NODE CLONING ATTACK IN WSNS

Node cloning attack is a communication attack. It is very harmful to a system which makes a decision based on the majority voting (Newsome et al. 2004). This is because it can generate a huge number of fake votes. Here first a node is captured by the attacker and then they extract its secret identity such as cryptographic keys, NodeID, and so on. By using the information collected, the attacker creates and installs a vast group of victim node clones at various network areas as shown in Figures 14.2 and 14.3. Such cloned nodes are now used to import a substantial percentage of bogus-sensed reports to the base station, which force it to make the wrong choice for many applications(Khan et al. 2013; Liu 2012; Chatterjee and Roy 2018c).

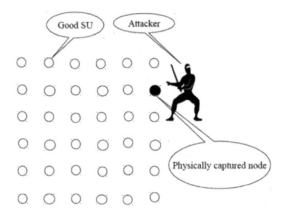

Figure 14.2   An attacker physically capture a node.

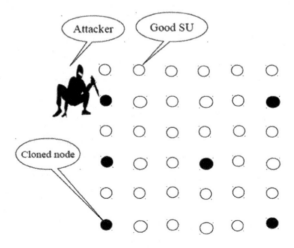

Figure 14.3   Clone of the capture node has been deployed.

A cloned node has many other bad impacts also on the network. A few of them are listed as follows:

- This type of attack has a very bad impact on the replication and fragmentation of data in the peer-to-peer and distributed storage system of WSN. When data is replicated or fragmented, it may be stored on a cloned node instead of the original node.
- The routing protocol of the wireless sensor network can be impacted heavily if the network contains cloned node's.
- The Network existence of clone nodes misleads the effects of data aggregation. A small number of malicious nodes false report can lead to wrong decision-making by the data aggregation system.
- A cloned node can also badly impact the network's resource allocation system. Because of the multiple cloned identities of the same node the proper allocation of the network resources is disturbed.

The balance of this survey will be complied with as follows: Section 14.2 represents countermeasures for node cloning attacks. Section 14.3 is the performance evaluation of node cloning attack detection techniques. Section 14.4 concludes this chapter.

14.2   COUNTERMEASURES FOR NODE CLONING ATTACKS

The techniques used to detect the node cloning attack can be broadly classified as centralized and distributed, as shown in Figure 14.4.

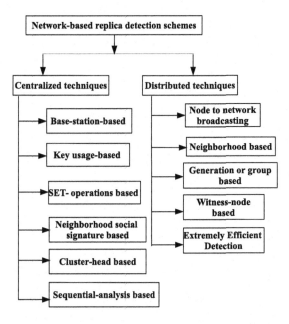

Figure 14.4   Classification of node cloning attacker detection techniques.

There is a strong central system called the base station in centralized techniques. All node-IDs and position argument are sent to the base station. Then, it will verify whether or not the same ID is present at more than one location. This system has very low overhead and a very high probability of detection. The disadvantage is that the entire system will be down if the central base station is down. The nodes closer to the base station use more power because of the heavy traffic that goes through them.

In distributed techniques, there is no concept of central decision-maker like the base station. A special approach is used, based on the witness-reporter. Neighbors' job here is to test a node's location argument. The advantage of this system is the absence of a single point failure. This system requires more resource to implement.

The node cloning attacker detection techniques in the CWSNs have been studied as follows.

14.2.1 CENTRALIZED TECHNIQUES

This technique is sub-categorized as follows:

1. **Base station based:** The authors suggested a compacted sensing-based replica identification (CSI) in (Yu et al. 2012). This technique is a centralized detection technique. Here, a predefined dataset is being sensed by each sensor nodes. Then, they broadcast it to its one-hop neighbors. The descendant node numbers are forwarded and aggregated around the aggregation tree using compressed sensing-based data collection methods. They considered the root of the aggregated tree as the central station. The central station gathers the consolidated information, and the data sensed by the network is recovered. If the sensory reading of a node is greater, then that node is the clone. It is because a node which is not cloned can only report the number once.

2. **Key-usage based:** Here we have the following two approaches:

   a. In (Chan et al. 2003), the authors have proposed a random key predistribution voting system (Eschenauer and Gligor 2002). When wrongdoing is noticed during the authentication process of the node to node, the voting scheme is initiated. The authors have proposed three different new techniques for a key establishment based on the predistribution of a set of random keys to each node.

      • q-composite keys: In network configuration, a secure connection is formed between any node and its neighbors by extracting from their key rings a single common key. The author increases the requirement of key overlap for key setup to provide better security to the network against node capture.

      • The multipath reinforcement: To create a secure link, they used to establish the link-key through multipaths.

      • The random-pairwise keys: Here, each node has the capability to check that few of its neighbors have certain secret keys. If yes, then they are legitimate nodes.

b. In (Brooks et al. 2007), the authors have proposed a protocol called random key predistribution (Eschenauer and Gligor 2002) for cloned element detection. The main idea of this technique is a random key predistribution which is used to assign the keys. A key is declared as clone when the usage of the key exceeds a threshold. Counting Bloom filters are used in this protocol to collect key usage statistics. The base station records how many times the network uses each key .

3. **SET-operation based:** In (Choi et al. 2007), the authors suggested an approach wherein the network is arbitrarily split into exclusive subsets. Each subset head collects subset member information. If there is no common portion of the network subsets, then there would be no clone attack. SET consists of five components:

   • First, the authors take the help of an ESMIS algorithm. The formation of the network subsets has been done in a distributedly using the ESMIS algorithm.

   • Second, the authors proposed the ESMIS algorithm along with an authentication scheme, which confirms the construction of the secure subset in a network consisting of compromised nodes.

   • Third, without compromising the security and the exclusiveness, the authors have optimized SET to form exclusive subset which has a randomization property.

   • Fourth, the authors have proposed a set computation technique on more than one trees. Intersection and union of subsets can be done efficiently on those trees.

   • Finally, the authors have proposed an interleaved authentication scheme, which preserves the reliability of set operation on a tree.

4. **Neighborhood-social-signature based:** In (Xing et al. 2008), the authors have proposed real-time attacker detection for WSN. The idea of the scheme is to allocate a fingerprint according to neighboring information. The fingerprint is determined with a superimposed s-disjunct script. Every node carries neighbor's fingerprints. Every node attaches that fingerprint when it sends out a message. Now the fingerprint can be verified by its neighbors. The method has two phases:

   • Computation of fingerprint: At the pre-deployment time, a superimposed s-disjunct code X is computed in the off-line mode. Here X is a binary matrix of order $N \times M$, where N is node count. Column of X is unique codeword of a node which is set previously.

   • Detection of clone attacks: Now, at network operation time, when u sends a message to v, v will verify the fingerprint to match with its records. One warning is lifted to the base station by v when a malfunction occurs.

5. **Cluster-head based:** Here, we have the following two approaches-
   a. In (Znaidi et al. 2009) the authors have presented a distributed hierarchical algorithm depending on the cluster head selection to spot node-replication attacks. Here, every cluster head uses a bloom filter to exchange the member's NODE-ID among each other to detect node replications.
   b. In (Chatterjee and Roy 2018a), the authors have proposed a Cuckoo filtering-based technique. Here, the cluster head collects the node-ids of its corresponding cluster members and try to find out whether there is any duplicate/repeated id present using Cuckoo filtering technique.

6. **Sequential-analysis based:** The authors have suggested in (Ho et al. 2009) a methodology based on Sequential Likelihood Ratio Check. Here, mobility speed of a node is used to detect a cloned node. The key idea of this scheme is to change the location of a cloned mobile node at a rate faster than the maximum velocity set by the system. Identifying the exact upper and lower speed limits is the main challenge here.

14.2.2 DISTRIBUTED TECHNIQUES

The distributed techniques are subcategorized as follows:

1. **Node-to-network broadcasting based:** In (Parno et al. 2005), the authors have proposed a technique for node-to-network broadcasting. Here the node's location claim is broadcast in the network by every node itself which is stored by its neighbors. When a conflicting location claim occurs, the replica is cut off by all its neighbors. This protocol has 100% clone node detection performance when we assume that the broadcasts reach every node in the network.
   In (Parno et al. 2005), the same authors have proposed a slightly different approach, which improves the performance of the previous algorithm. The algorithm deterministically chooses a subset of witness nodes. It shares a claim for position of nodes with only certain restricted subsets. The authors have chosen a function of NODE-IDs to select the witnesses. When there is a duplicate of the node, the witnesses will find different locations of the same node. This algorithm is called as Deterministic Multicast.
   To further improve the performance of the Deterministic Multicast, two other algorithms have been proposed, namely, Randomized-Multicast(RM) and Line-Selected-Multicast(LSM). The RM randoAll node-IDs and position argument are sent to the base station mly selects a set of witness-nodes and distributes location claims among this set. The LSM takes the help of the routing-topology of the network for selecting witnesses-node.

2. **Neighborhood-based:** Here, we have the following two approaches-

   a. In (Liu 2012), the authors have proposed the single hop detection (SHD). The principal concept of the SHD system is now a physical node (i.e., a node's ID and private key) cannot be found in the neighborhood list of different nodes at the same time. In this protocol, the neighbors refer to the list of the one-hop neighbor of a node. The SHD protocol involves two stages:

    i. The fingerprint-claim phase: Here discrete node joins its neighboring one-hop list. The fingerprint claim of the node is being forwarded to its one-hop neighborhood. While the neighbor node gets the request for a fingerprint, it determines whether or not it will watch the alleged node. If decision is yes, then it will test the fingerprint-claim.

    ii. The fingerprint-verification phase: Let the list of witnessed nodes wl_i and wl_j is swapped between node i and j through the piggybacking of Hello message. If, $wl_i \cap wl_j = \emptyset$, then the further exchange of the fingerprint claims takes place between the nodes in $wl_i \cap wl_j$. Then, they check for a claimed conflict. Any conflict implies replica exists in the network.

  b. In (Wang et al. 2017), the authors have proposed a distributed, hybrid clone node detection scheme. The proposed scheme consists of local and global detection phase. A small part of the network is considered as the local area on which the local detection is performed. This improves the meeting probability of cloned nodes. If the cloned nodes are not closed to each other, then global detection can be used to capture the malicious node.

3. **Generation or group based:** In (Bekara and Maknavicius 2007; Bekara and Maknavicius 2012), the authors have proposed a generation-based or group-based techniques. A symmetric polynomial is used to fix a key between each pair of nodes to limit the order of deployment and defined a group-based deployment model. In this technique, the sensors are deployed continuously in consecutive generations. In this technique, each node is from a specific generation. Here the chance to fix a pairwise key with the neighbors is given only to the new entry. Every network node knows the maximum number of generations deployed. In such scenario, there will be no chance for the cloned nodes to fixed a pair-wise key as because they belong to an older deployed generation.

4. **Witness-Node Based:** In (Ming et al. 2009), the authors have proposed a technique based on bloom filter. This technique uses cell forwarding and cross forwarding. Their technique is divided into four new memory and power-efficient applications as follows:

  • **B-MEM:** B-MEM reduces the number of position statements which are stored in the node. To do this, it uses two lightweight bloom filters. To detect clone nodes, the position claim information is stored in the bloom filter. They proposed a cell-forwarding method of solving the cross-over problem and suggested the algorithms below.

  • **BC-MEM:** This is used to increase the likelihood of identification while further reducing overhead memory.

- **C-MEM:** It is used to focus on another problem called the crowded centre problem. It uses another recent method called cross forwarding that evenly distributes overhead communication and memory overheads across all the network nodes.
- **CC-MEM:** This protocol combines cell and cross forwarding to achieve the maximum efficiency.

5. **Extremely Efficient Detection:** In (Yu et al. 2008), the authors have suggested the "Extremely Efficient Detection (XED)"in a localized manner. This algorithm exchanges some random number as code among two nodes. Every time the nodes recognized themselves by that code. A random number is generated with some cryptographic hash function. More memory is required to store the authentication key of the nodes.

14.3 PERFORMANCE OF NODE CLONING ATTACK DETECTION SCHEMES

Table 14.1 contains a summary of the complexity of different centralized node cloning attack detection techniques. Where n represents the node count, Ç present messages from the sensor node, T is the pillar value in the S-disjunct superimposed

Table 14.1

Asymptotic Complexity of Centralized Node Cloning Attack Detection Technique

Method Used	Type of Network	Communication Overhead	Storage Overhead
Base station-based (Yu et al. 2012)	Static	$O(nlogn)$	$O(nlogn)$
Key usage-based (Chan et al. 2003)	Static and Mobile	$O(nlogn)$	$O(n)$
Key usage-based (Brooks et al. 2007)	Static and Mobile	$O(nlogn)$	$O(n)$
SET operation-based (Choi et al. 2007)	Static	$O(n)$	$O(\delta)$
Neighborhood social signature-based (Xing et al. 2008)	Static	$O(Ç.(1+r))$	$O((\delta) + min(\kappa, T.log_2\kappa))$
Cluster head-based (Znaidi et al. 2009)	Static	$O(n^2)$	$O(n)$
Cluster head-based (Chatterjee and Roy 2018a)	Static	$O(n)$	$O(n)$
Sequential analysis-based (Ho et al. 2009)	Mobile	$O(n\sqrt{n})$	$O(n)$

Table 14.2

Asymptotic Complexity of Distributed Node Cloning Attack Detection Technique

Method Used	Type of Networks	Communication Overhead	Storage Overhead
Node-to- Network Broadcasting based (Parno et al. 2005)	Static	$O(n^2)$	$O(1)$
Neighbor hood based (Liu 2012)	Static	$O(r\sqrt{n})$	$O(r)$
Neighbor hood based (Wang et al. 2017)	Static	$O(d^2f)$	$O(d^2f)$
Generation or group based (Bekara and Maknavicius 2007; Bekara and Maknavicius 2012)	Static	$O(\sqrt{n})$	$O(1)$
Witness-node based (Ming et al. 2009)	Static	$O(kn\sqrt{n})$	B-MEM: $O(t\kappa+t'\kappa\sqrt{n})$
			BC-MEM: $O(t\kappa+t'\kappa\sqrt{n'})$ C-MEM: $O(t+t'\sqrt{n})$ CC-MEM: $O(t+t'n')$
Extremely Efficient Detection (Yu et al. 2008)	Mobile	$O(1)$	$O(n)$

script, δ is the degree of adjacent nodes, κ is row count in S-disjunct superimposed code, $r = log_2\kappa/packet - length \times 100\%$ and packet-length is the bit-length of a normal massage.

Table 14.2 contains a summary of the complexity of different distributed node cloning attack detection techniques. Where n is network node count, r is the communication radius, average number of line segments per argument is κ, T is the volume of a position statement and t' presents the byte count used by a bloom filter to register the affiliation about an entity, d is the number of neighbors, and f is the quantity of position claim slots.

Table 14.3 contains a summary of the advantages and disadvantages of different clone node detection schemes.

14.3.1 DISCUSSION

The node cloning attack itself has a very bad impact on the network. In the presence of these attacks, the cloned node sends false sensing information in large numbers to

Table 14.3
Advantages and Disadvantages of Node Cloning Attack Detection Techniques

	Centralized Techniques	
Method Used	**(+)Advantages**	**(−)Disadvantages**
Based-station based (Yu et al. 2012)	Strong probability of identification.	High communication and storage cost.
Key-usage based (Chan et al. 2003)	NA	Very less detection probability.
Key-usage based (Brooks et al. 2007)	NA	Higher rate of false positive and false negative.
SET-operation based (Choi et al. 2007)	Not depends on node-position, needs less memory.	Single point failure.
Neighborhood social signature based (Xing et al. 2008)	Less computation cost.	Depends on the fingerprint provided by the neighbors.
Cluster-head based (Znaidi et al. 2009)	Low communication overhead.	Low detection probability.
Cluster-head based (Chatterjee and Roy 2018a)	Low communication overhead.	Not suitable for very big network.
Sequential analysis based (Ho et al. 2009)	Strong probability of identification.	Upper and lower speed limit of a node is not fixed.
	Distributed Techniques	
Method used	**(+)Advantages**	**(-)Disadvantages**
Node-to-Network Broadcasting-based (Parno et al. 2005)	Detection rate is high.	Communication cost is high.
Neighborhood based (Liu 2012)	Not depends on location.	High communication cost.
Neighborhood based (Wang et al. 2017)	Not depends on location.	High communication cost.
Generation or group based (Bekara and Maknavicius 2007; Bekara and Maknavicius 2012)	Needs less communication overhead.	Nodes are confined within it,s group.
Witness-node based (Ming et al. 2009)	Detection probability is high and less memory requirements.	Depends on node position.
Extremely Efficient Detection (Yu et al. 2008)	High detection probability.	Needs more memory.

the collaborative decision-making system of an IoT application. As a result, the collaborative system makes the wrong decision, and the IoT application fails to makes some decisions. Effective countermeasures against this attack are very important as a whole for the proper implementation of different network functions in WSN-based IoT applications.

14.4 CONCLUSION

In this chapter, we have systematically reviewed the different state-of-the-art cloned node attacker detection techniques applicable for WSN-based IoT applications. The current techniques are divided into two groups: centralized and distributed, and their sub-categories are then listed. The techniques for cloned node detection are proficient in many respects, but at the same time, they have some drawbacks also. In conclusion, however, the cloned node attacker detection techniques still have several issues and challenges to mitigate in order to become more appropriate for real-life IoT applications and acceptable for resource-constrained WSNs .

REFERENCES

Akan, O., B., Karli, O., B., and Ergul, O.(2009) Cognitive radio sensor networks, *IEEE Network: The Magazine of Global Internetworking - Special issue title on networking over multi-hop cognitive networks*, Vol. 23, No. 4, pp 34–40.

Brooks, R., Govindaraju, P., Y., Pirretti, M., Vijaykrishnan, N. and Kandemir, M., T. (2007) On the detection of clones in sensor networks using random key predistribution. In *Proc. of the IEEE Transactions on Systems, Man and Cybernetics* , pp. 1246–1258.

Bekara, C. and Maknavicius, M., L.(2007) A new protocol for securing wireless sensor networks against nodes replication attacks. In *Proc. of the 3rd IEEE International Conference on Wireless and Mobile Computing, Networking and Communications (WiMob 07)*.

Bekara, C. and Maknavicius, M., L. (2012) *Defending against nodes replication attacks on wireless sensor networks.*.

Chatterjee, P., S. and Roy, M. (2015) A Regression based Spectrum-Sensing Data-Falsification Attack Detection Technique in CWSN. In *14th International Conference on Information Technology*, pp. 48–53.

Chan, H., Perrig, A. and Song, D., X.(2003) Random key predistribution schemes for sensor networks. In *Proc. of the IEEE Symposium on Security and Privacy, IEEE Computer Society* , pp. 197–213.

Choi, H., Zhu, S. and Porta, T., F., L. (2007) Set: detecting node clones in sensor networks. In *Proc. of the 3rd International Conference on Security and Privacy in Communication Networks (SecureComm 07)*, pp. 341–350.

Chatterjee, P., S. and Roy, M. (2017) Using a UNPCC based classification algorithm for detection of PUE attackers in cognitive wireless sensor networks. In *Proc. of IEEE International Conference on Advanced Networks and Telecommunications Systems (ANTS)*, IEEE.

Chatterjee, P., S. and Roy, M. (2018) Lightweight clone-node detection algorithm for efficiently handling SSDF attacks and facilitating secure spectrum allocation in CWSNs. In *proc. of IET Wireless Sensor Systems*, IET Publication.

Chatterjee, P., S. and Roy, M. (2018) Detecting PUE attack by measuring aberrational node behavior in CWSN. In *Proc. of the Journal of Interconnection Networks*, World Scientific Publication.

Chatterjee, P., S. and Roy, M. (2018) Maximum match filtering algorithm to defend spectrum-sensing data falsification attack in CWSN. In *Proc. of the International Journal of Wireless and Mobile Computing*, Inderscience Publication.

Eschenauer, L. and Gligor, V., D. (2002) A key-management scheme for distributed sensor networks. In *Proc. of the 9th ACM Conference on Computer and Communications Security*, pp. 41–47.

Haykin, S. (2005) Cognitive radio: Brain-empowered wireless communications. *IEEE Journal on Selected Areas in Communications (JSAC)*, Vol.23, No. 2, pp. 201–220.

Ho, J., W., Wright, M. and Das, S., K. (2009) Fast detection of replica node attacks in mobile sensor networks using sequential analysis. In *Proc. of the INFOCOM*.

Khan, W., Z., Aalsalem, M., Y. and Xiang, Y. (2013) Detection and mitigation of node replication attacks in wireless sensor networks: A survey. In *Proc. of the International Journal of Distributed Sensor Networks*, Hindawi Publishing Corporation.

Liu, E. (2012) Single hop detection of node clone attacks in mobile wireless sensor networks. In *Proc. of the International Workshop on Information and Electronics Engineering (IWIEE)*, pages 2798–2803.

Mitola, J. (2000) *Cognitive radio: An integrated agent architecture for software defined radio.*, PhD thesis, Royal Institute of Technology (KTH), Stockholm, Sweden.

Ming, Z., Vishal, K., Shigang, C. and Xuelian, X. (2009) Memory efficient protocols for detecting node replication attacks in wireless sensor networks. In *Proc. of the 16th IEEE Intl. Conf. on Network Protocols (ICNP09)*.

Newsome, J., Shi, E., Song, D. and Perrig, A. (2004) The sybil attack in sensor networks: Analysis and defenses. In *Proc. of the Research Showcase @ CMU*, The Carnegie Institute of Technology.

Parno, B., Perrig, A. and Gligor, V. (2005) Distributed detection of node replication attacks in sensor networks. In *Proc. of the IEEE Symposiumon Security and Privacy (IEEE S and P 05)*, pp. 49–63.

Wang, Z., Zhou, C. and Liu, Y. (2017) Efficient hybrid detection of node replication attacks in mobile sensor networks. In *Proc. of Mobile Information Systems*, Hindawi.

Xing, K., Cheng, X., Liu, F and Du, D., H., C. (2008) Real-time detection of clone attacks in wireless sensor networks. In *Proc. of the 28th International Conference on Distributed Computing Systems (ICDCS 08)*, pp. 310.

Yu, C., M., Lu, C. S. and Kuo, S., Y. (2012) Csi: compressed sensing based clone identification in sensor networks. In *Proc. of the IEEE International Conference on Pervasive Computing and Communications Workshops (PERCOM Workshops 12)*.

Yu, C., M., Lu, C., S. and Kuo, S., Y. (2008) Mobile sensor network resilient against node replication attacks. In *Proc. of the 5th annual IEEE Communications Society Conference on Sensor, Mesh and Ad Hoc Communications and Networks (SECON)*.

Znaidi, W., Minier, M. and Ubeda, S.(2009) Hierarchical node replication attacks detection in wireless sensors networks. In *Proc. of the 20th IEEE Personal, Indoor and Mobile Radio Communications Symposium (PIMRC 09)*.

Index

Agobot 232
AI 201
Amazon Web Services (AWS) 150
ARPANET 2
Artificial neural networks (ANN) 112, 200

BMAP 37
Botnet 8, 232

(C&C) channel 232
Convolutional Neural Network (CNN) 152
COVID-19 205
CRU 7
cyber physical system 164

deep learning (DL) 131
deep neural network (DNN) 113
Denial Distributed of Service (DDoS) 7, 230
DLBS 69

ELBS 69
electronic health records (EHRs) 198
ETX networks 212
Extremely Efficient Detection (XED) 256

FFTF 70
fog computing 68
4G-LTE 91, 149

gross value added (GVA) 119

horizontal axis wind turbine (HAWT) 93
HTTP Flood 231

ICT 91
Industrial IoT (IIoT) 1
Internet control message protocol (ICMP) 231
Internet of Healthcare Things (IoHT) 179
Internet of Nano Things (IoNT) 178
Internet of things (IoT) 1, 15, 35, 87, 106, 119
intrusion detection system (IDS) 212

IPv4 5
IPv6 6
IRC 235
ISO/OSI protocol stack 231

LIGA 21

machine learning (ML) 106
malware 8
MEMS 20
microelectronics 20
micromachining 20
multilayer perceptron (MLP) 158

NFC, ZIgBee, Bluetooth 220
node cloning 250

OCR technology 167

principle of component analysis (PCA) 152

Quality of Service (QoS) 177

radial basis function (RBF) network 152
RBot 232
RVNS 69

SDbot, Spybot 232
SDN 69
SEaaS 36
SFC 69
SLA 222
smart sensor 17
SSID 28
SVELTE technique 212

VFN 74

WAN/LAN 95
Wi-Fi, Bluetooth, LoWPAN, RFID 198, 216
WRR 80
WSN 248

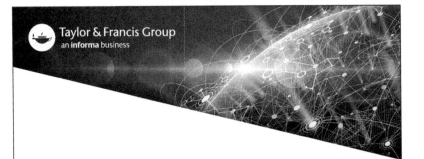

Printed in the United States
by Baker & Taylor Publisher Services